THE FUTURE OF SMALL TELESCOPES IN THE NEW MILLENNIUM

ASTROPHYSICS AND SPACE SCIENCE LIBRARY

VOLUME 287

THE FUTURE OF SMALL TELESCOPES IN THE NEW MILLENNIUM

Volume I – Perceptions, Productivities, and Policies

Edited by

TERRY D. OSWALT

Florida Institute of Technology,
Melbourne, Florida, U.S.A.

KLUWER ACADEMIC PUBLISHERS

DORDRECHT / BOSTON / LONDON

A C.I.P. Catalogue record for this book is available from the Library of Congress.

QB
88
.F993
2003
v. 1

ISBN 1-4020-0948-8 (HB)
ISBN 1-4020-0951-8 (Indivisible set)

Published by Kluwer Academic Publishers,
P.O. Box 17, 3300 AA Dordrecht, The Netherlands.

Sold and distributed in North, Central and South America
by Kluwer Academic Publishers,
101 Philip Drive, Norwell, MA 02061, U.S.A.

In all other countries, sold and distributed
by Kluwer Academic Publishers,
P.O. Box 322, 3300 AH Dordrecht, The Netherlands.

Printed on acid-free paper

Printed in the Netherlands.

Contents

Preface

The recent closures of telescopes smaller than 4-m in aperture are not the impetus for this book. They are however the most visible sign of pressure faced by all small facilities. The fact is that the future of these facilities may be quite bright if creative steps are taken. Some of these steps should parallel the effort put into the development and operation of the new generation of 8-m class telescopes. However a number of innovative approaches are practical only at smaller facilities.

The seed of this book germinated at the Lowell Workshop on the "Future of Small Telescopes" in November 1996. That meeting coincided with the second major closure of a telescope at Kitt Peak National Observatory. Many astronomers were understandably concerned about what the next few years might bring for telescopes smaller than ~4-m. Most felt that small facilities were being overlooked as a sort of "aperture envy" spurred the rush to build new 8- to 10-m class public and private telescopes. The time was ripe for the astronomical community to compare notes on what essential roles were really being played by small telescopes. The meeting identified many areas of intense research activity and provided the incentive to organize several follow-up sessions at AAS meetings in 1997-2000. All these recent meetings tended to focus on cutting edge science currently being done, rather than on building a community consensus on future priorities.

The 2000 Atlanta AAS meeting featured a special session on the future of small telescopes. Several of the authors in this volume presented papers at that session. At this session I was invited by Kluwer to edit a book on the future of small telescopes. This was an opportune time, as the U.S. Decade Survey was being organized and it seemed clear from early reports that it would focus on the most expensive next-generation large projects that only federal funding agencies or large private consortia would be able to fund.

Astronomy is one of the few sciences that regularly undertake a Decade Survey. We receive strong support from the government because of our ability to prioritize our science initiatives and funding needs. However, many who attended the Atlanta AAS meeting felt that the directives under which our Decade Study is conducted leave out the cutting edge "big science" that can only be done with small facilities. Also conspicuously overlooked is the essential education, instrumentation and research support they provide to 8-m class telescopes.

This project is one of the first attempts to fill a serious void in planning for the next decade of astronomical research. The problem is international in scope, as many of the articles in these volumes will attest. A wide cross section of the astronomical community was invited to help visualize and prioritize the next decade of small telescope science. A surprisingly large fraction of those invited accepted the challenge of helping us accomplish this goal—large enough to fill several volumes. Many more ideas were offered by prospective authors than could be contained in one volume—strong testament to the broad support that small telescopes have among our scientific colleagues.

Most of the following chapters start with what work is currently underway and build towards recommendations for what should be done in this new decade. Not surprisingly, a wide variety of opinions are represented in the following pages, yet the consensus on the essential role of small telescopes is unequivocal. Even so, only the surface of this topic has been scratched. We hope that enough facts are presented not only to successfully press the case for continued investments in small public and private facilitates, but especially to spur planning committees and funding agencies to include small telescopes in the science initiatives recommended by the Decade Survey. It is our fervent hope that the astronomical community will undertake many of the ideas described in these volumes in the coming decade.

This is a compendium of examples and recommendations. Beginning with the astronomical community's own perceptions of small telescope productivity, we examine the most likely technology that can upgrade existing facilities and build the next generation of small telescopes. Although the main focus is on ground-based optical telescopes—this being the largest current area of small telescope use—these volumes include the smaller cost-effective radio and space-based facilities that face similar problems in prioritization and funding, yet provide unique opportunities for discovery during the next decade.

Much can be gained simply by improving the operational efficiency of existing facilities. This strategy begins with better use of standard off-the-shelf technology and optimization of individual telescopes for special projects. It extends to include active and adaptive optical improvements and beyond, to new modes of scheduling, automation, and Internet access.

We also explore a selection of scientific initiatives that are ideally suited for small telescopes. This spans the entire range of astronomical research, from solar system astronomy to cosmology. Hopefully, many of these articles will provide fundamental references for those looking to undertake new projects and especially for those who are called upon by funding agencies and/or policy making committees to set scientific priorities for the coming decade.

Many topics have been left out for lack of space. In many cases prospective contributors were simply unable to fit a chapter assignment into their already busy schedules. The overwhelming majority expressed strong enthusiasm and support for this project. Among the most obvious areas that are missing from these volumes is a perspective from the U.S. national observatories. Several potential authors were invited. However, budget cuts and advisory committee recommendations have narrowed the mission of U.S. national observatories to essentially the operation of existing 4- to 8-m class instruments and the development of the next generation ~30-m optical/infrared and 8-m class special purpose telescopes[1]. In fact, even 4-m telescopes are at risk of being overshadowed by this new generation of telescopes during the next decade. Most of the smaller telescopes at these sites have been or will be "privatized" or closed during the coming decade. Thus, it is unlikely that the national facilities will play a major role in setting priorities for them in the future.

The closures of small telescopes at national facilities are not unique to the U.S. There is a common misconception that all 1-2-m telescopes can be built and supported by universities, and that even 4-m class telescopes should be run by university consortia. Certainly this is part of the future. However, we hope that this book will help underscore the fact that such a simplistic strategy will lead to a much less diverse research activity and that a suite of publicly-available international facilities and instruments need to be supported.

We need to address the question of how the operations of small facilities are to be supported. Clearly, the future calls for more inventive ways of funding both large and small facilities. The current trend towards all telescopes of aperture ≤4-m being supported by universities and consortia is shortsighted. Nevertheless, these volumes are not about how to "save" small public and private astronomical facilities. Rather, we focus on setting appropriate scientific priorities and identifying the best means of achieving them. Hopefully, presented herein is a factual basis from which workable solutions can be found. Whatever the solutions, they should not just preserve productive existing facilities, but also provide for the planning and construction of the next generation of small ground- and space-based, optical and non-optical facilities of all types.

Clearly, if we take the opportunity now to include small facilities in our national strategic plan, the scientific return on our investments in 8-m class

---

[1] See the article by Castelaz (1999 IAPPP Communications 75, 67) for a tongue-in-cheek, yet sobering, discussion of the consequences of headlong pursuit of the "ultimate telescope."

instruments and science as a whole will be enormously increased. Without our input at this important juncture, astronomical research will be limited to a much smaller range of topics, a much smaller fraction of the community, and a much smaller rate of discovery. The case is clear: there is much science left that can only be done with small telescopes!

I would like to thank all our contributing authors for helping us take this important step into the future. A special note of gratitude is due Harry Blom, our Kluwer editor. Without his constant encouragement and patience these volumes would never have become a reality. I would also like to thank Victoria Astley, Tamalyn Heinz, Kyle Johnston, Kandy Kasprick, and Stepfanie Rafferty, who helped in the difficult and tedious preparation of the many pages of manuscripts. Most of all, I would like to thank my wife Barbara for putting up with the long hours I spent on this project. To her these volumes are dedicated.

It is truly an exciting time for small telescope users. We are challenged to plan the future of the true workhorses of astronomical research. Won't you join us?

Terry D. Oswalt
Florida Institute of Technology
Melbourne, Florida
April 2002

Chapter 1

Small Telescopes in the New Millennium

John Huchra
Harvard-Smithsonian Center for Astrophysics
Cambridge, Massachusetts USA

Abstract: As we enter the new millennium, astronomers have or will soon have over fifteen 8-meter class telescopes to work with. However, there is likely to remain a need for small telescopes, especially specialized telescopes that can be optimized for and dedicated to specific task. These can be well justified scientifically for projects ranging from solar system studies to cosmology. Small telescopes also are the heart and soul of teaching and public outreach programs. One thing is changing, however, and that is the definition of "small."

Key words: small telescopes, science priorities, astronomy education

1.   INTRODUCTION

The world astronomical community is emerging into the 21st century at the end of a decade devoted to the construction of a new generation of very large telescopes. This telescope-building boom was driven by the continued scientific need to press deeper and fainter and by the increased demand for optical and infrared astronomical observations to support space missions and to follow up on spectacular discoveries in cosmology, star and planet formation, and the general evolution of the contents of the Universe. The drive to larger, better telescopes was also driven by the possibility of improvements in imaging performance and followed the development in the 1980's of almost perfect electronic detectors in the optical and near-infrared. Modern low noise, large format array detectors based on Silicon, Indium-Antimonide and Mercury-Cadmium-Telluride together cover the full range of OIR wavelengths accessible from the ground, 0.3-5 microns, with nearly

T.D. Oswalt (ed.), The Future of Small Telescopes in the New Millennium, Vol. I, 1–5.
© 2003 *Kluwer Academic Publishers. Printed in the Netherlands.*

100% quantum efficiency and very low readout noise and dark current). The only way to go was up.

Even as larger and larger telescopes are contemplated for their increased collecting area and increased angular resolution, there still is a need for small telescopes in our research and teaching arsenal. As very large telescopes become both more common and more specialized, the demand for time on them has concomitantly increased. Astronomy is a growing field. In the US alone, the size of the professional astronomical community grew by nearly 50. There has been significant growth in the number of colleges and universities with astronomy departments or with significant astronomy components in their physics departments.

A second consequence of the improvement in detectors has been that even "small" telescopes are capable of making forefront observations and obtaining forefront datasets. The most current examples in the US include the wealth of data coming from the 2 Micron All-Sky Survey (two 1.3m telescopes), the Sloan Digital Sky Survey (a wide field 2.5m), the study of rapid bursts with ROTSE 0.45m, supernovae searches with the KAIT 0.76-m telescope, and the search for transients with the LOTIS 4x0.111m cameras, mapping ionized gas in the northern sky with WHAM, the 0.6m Wisconsin H-alpha Mapper, and searches for killer asteroids with the 1m LINEAR telescope.

Lastly, amateur astronomers, often in clubs, are capable of building 1m class telescopes and participating in real astrophysical research such as variable star observations or supernovae searches, and also in the incredibly important arena of public outreach. The same is true for small colleges, where students, using small telescopes and commercially available, intermediate format electronic cameras, can contribute to fundamental research on subjects as varied as the properties of stars to the measurement of distances in the Universe (See almost any issue of the journal of the American Association of Variable Star Observers).

Do we continue to need small telescopes?

2. DRIVERS

There are two simple yet important drivers for the continued support of small telescopes:

2.1 The Right Tool For The Job

Many of the programs in the above paragraph require specialized telescopes, either because of constraints on time and access, field-of-view, survey work or slew characteristics. Monitoring variable objects, especially if full phase coverage is essential, requires telescopes with stable instrumentation and very flexible schedules. Every large telescope currently in operation has a suite of instruments that changes with the phase of the moon—usually a faint object spectrograph and camera when the sky is dark and a high-resolution spectrograph and/or an IR camera and spectrograph for when the moon is up. For very broad scientific programs, such a suite optimizes the scientific output of the observatory. However, there are key programs, such as following SN light curves (Riess et al 1999) or hunting for Machos (Alcock et al 2001) or searches for variable stars (DIRECT, Macri et al. 2001) where the optimal strategy is to have a single instrument, a CCD camera, on the telescope at all times.

Similarly, for large area surveys that do not require the deepest possible imaging, the aerial coverage of a small telescope varies inversely as the square of the diameter of the telescope for fixed focal ratio. Since modern, state of the art, large area detectors can cost as much as a small telescope (the going rate last year for a 2048x2048 pixel HgCdTe array was nearly a quarter of a million dollars), paving a large angular extent on a small telescope is almost two orders of magnitude cheaper on a 1m than an 8m. Classic examples again are 2MASS and WHAM (Reynolds et al 1998). Generally, the size of a telescope needed for imaging surveys to provide object identifications for spectroscopy goes as the square-root of the relative spectral resolution; in the case of imaging through normal filters (R = 5) relative to classification spectroscopy or that needed for galaxy radial velocity dispersions (R=500), the number is $\sim(500/5)^{1/2} = 10$, so a 1m telescope can serve effectively as a finder for a 10-m assuming other site and telescope characteristics are equal. This argument is the same as that used well over half a century ago to justify the building of the 1.2-m Palomar Schmidt telescope to serve as a "finder" for the 200-inch Hale telescope, and is still valid today. (Note: To be fully correct, one needs to modify the above formula to take into account the different spatial resolutions and noise sources to estimate the proper ratio for any specific case. For example, if the diffraction limit, rather than the natural site seeing limit, is reached at both telescopes, a somewhat large imaging telescope is required.) Also rather curiously, if a bright supernova went off in our Milky Way or one of our nearby neighbor galaxies in the Local Group, we would need a good spectrograph (like the ones at the 1.5m telescopes at Mt. Hopkins and Cerro Tololo) to follow it up—the instruments on the very large telescopes would saturate in less than the minimum allowable integration time!

Finally, in the tool department, there are projects like survey spectroscopy of bright galaxies where Bowen's law (Bowen 1964, Case B) for moderately extended objects holds. That is to say the gain in signal to-noise goes only linearly with telescope diameter. If small telescopes can use multi-object spectrographs, such as the 120-fiber spectrograph on the 1.2m SRC Schmidt telescope, currently being used for the 6dF survey, the bang for the buck can be enormous. Again, the match of a high quality instrument with a specific scientific objective (in this case the southern 2MASS Redshift Survey, Huchra 2001) yields a winning strategy with a very small telescope.

2.2   Education, Training and Public Outreach

Despite the promise of the web and "Virtual" observatories, there is still no substitute for getting kids to telescopes. Here kids can be defined as primary and secondary school students, undergraduates, graduate students, especially those interested in instrumentation, and anyone else interested in seeing the sky up close and personal (kids at heart!). As of June 2002, there were almost 260,000 amateur astronomy sites listed on the web and over three million astronomy sites overall. Of all the sciences, there is little doubt that astronomy has the greatest public appeal and almost all of that is positive (in contrast to nuclear physics, which has a rather mixed public reputation). Of all the physical sciences, astronomy is the only one with a host of its own public and private "museums"—planetaria. The National Air and Space Museum is the Smithsonian's biggest draw.

Astronomy is also one of the best hooks for getting college students interested in science. Statistics compiled by the AIP and the AAPT indicate that more undergraduates take general astronomy than just about any other science course. Astronomy problems are usually very visual and can be used to convey a number of very fundamental concepts. Using telescopes to do such problems, by eye at a general level and with low-cost electronic cameras for more advanced courses, is rated over and over again as one of the best learning experiences for college students. In fact, since there is so much real science that can be done with small telescopes, it is one of the best ways of involving advanced undergraduates in research. The number of small to intermediate (0.6-1.0m) "teaching" telescopes is astronomical, and it is not expensive to provide them with very reasonable instrumentation for even more advanced projects (Kannappan et al. 2002). For example, I'm now involved in a project to provide a local secondary school with a well-instrumented 24-inch telescope!

And at higher levels, small telescopes provide opportunities for graduate students to learn first hand about instrumentation, often by being involved in building or operating new, simple instruments—I still remember working with Jim Gunn to implement a new image intensified camera for the Palomar 60-inch when I was a student. While at the National Observatories and at most major private observatories students are neither encouraged nor often allowed to experiment with instrumentation on 8m class telescopes, building simple instruments for small telescopes is one of the best ways of training students (Barden, Ramsey & Truax 1980; Claver 1992), including training them in developing and leading the construction of instrumentation.

3.   SUMMARY

There is clearly a need for small telescopes, even in the era of 8m behemoths and planning for 20, 30 50 and 100m giant and "overwhelmingly large" telescopes.   This need is driven by two important tasks: (1) fundamental research most economically and efficiently done on small telescopes, and (2) education, training and public outreach. For research, the definition of small is evolving. It still includes the 1m class telescopes and smaller for specific tasks, but now also includes survey telescopes like the SDSS 2.5m.   In almost all such cases, telescopes and projects like this deserve, nay, required national funding since their products are of and in the national interest.

For teaching, it is hard to beat small telescopes and the now low cost associated instrumentation for getting students and the public involved.   In these cases, support for such telescopes is in the purview of the institution—school, college, university, planetarium—that operates them.

Whatever the future holds for giant telescopes and the science that drives them, cost effective, well instrumented, small telescopes have their place in the arsenal of science. The author would like to thank all his colleagues in crime—doing cosmology with small telescopes—over the last three decades. This paper was written at the Aspen Center for Physics, which is supported by the NSF and NASA.

4.   REFERENCES

Alcock, C. et al 2001, ApJL 550, L169.

Barden, S. Ramsey, L. & Truax, R. 1980, BAAS 12, 460.

Bowen, I. 1964, AJ 69, 816.

Claver, C. 1992, BAAS 24, 1282

Federal Funding of Astronomical Research, a report of the NRC, 2000.

Huchra, J. 2000, in *Cosmic Flows 1999*: Towards an *Understanding of Large-Scale Structure,* S. Courteau, M. Strauss & J. Willick, eds., PASP Conf. Series Vol 201, 96.

Kannappen, S., Fabricant, D. & Hughes, C. 2002. PASP 114, 577.

Macri, L., Stanek, K., Sasselov, D., Krockenberger, M. & Kaluzny, J. 2001, AJ 121, 870.

Reynolds, R. J. et al. 1998, PASA 15, 14.

Riess, A. et al. 1999, AJ 117, 707.

Chapter 2

The Decade Survey in Astronomy and Astrophysics

Christopher F. McKee
University of California
Berkeley, California USA

Joseph H. Taylor
Princeton University
Princeton, New Jersey USA

Abstract: The Astronomy and Astrophysics Survey Committee has recommended an ambitious program of ground-based and space-based initiatives to enable continued exploration of the universe in the coming decade. The top recommendation is the Next Generation Space Telescope, which will study the dawn of the modern universe. Small telescopes will make key contributions to the surveys that will be archived in the National Virtual Observatory, the top-ranked small initiative. In addition to their research role, small telescopes are essential for education and for training the next generation of instrumentalists.

Key words: Astronomy and Astrophysics Survey Committee

We live in a golden age of astronomy and astrophysics. In the last decade alone a number of major discoveries were made. These include planets orbiting other stars; the Kuiper belt of small, primitive bodies in the outer solar system; "brown dwarfs," stars too cool to sustain nuclear reactions in their interiors; the phenomenon of gravitational microlensing; the cosmological nature of gamma ray bursts; massive black holes in the nuclei of galaxies; the theoretically predicted fluctuations in the cosmic microwave background radiation; and the remarkable result that the expansion of our universe is accelerating, suggesting the presence of "dark energy." These discoveries were made with telescopes of all sizes, covering a broad range of wavelengths.

The coming decade offers the promise of discoveries at least as important as those of the previous one. Furthermore, in addition to adding to our factual knowledge of the universe, we have the prospect of significantly

T.D. Oswalt (ed.), The Future of Small Telescopes in the New Millennium, Vol. I, 7–11.
© 2003 *Kluwer Academic Publishers. Printed in the Netherlands.*

adding to our *understanding* of the universe. To do this will require a new generation of telescopes and instruments, both on the ground and in space. The community of astronomers and astrophysicists is unique among scientific communities in the United States in that each decade a committee is established to assess the state of the field and recommend priorities for the coming decade.

The most recent Astronomy and Astrophysics Survey Committee was charged in 1998 with setting priorities for the first decade of the new millennium. This committee consisted of 15 distinguished astronomers and astrophysicists from a variety of institutions. Seven disciplinary sub-panels provided advice to the committee on particle, nuclear, and gravitational wave astrophysics; radio and submillimeter astronomy; optical and infrared astronomy from the ground; ultraviolet, optical and infrared astronomy from space; high energy astrophysics from space; solar astronomy; and theory, computation and data exploration. In addition, one panel focused on astronomy education and policy, while another panel assessed the benefits to the nation from astronomy and astrophysics. Each panel considered many suggestions for new initiatives for the coming decade, and members of the panels used their scientific and technical judgement to select and promote the very best of these. These concepts were then brought forward to the AASC, which winnowed the list further and prioritized the projects and initiatives that survived. Great care was taken to have a thorough and thoughtful discussion of each proposal, so that the final recommendations could reflect a true consensus of the entire committee.

Setting priorities in a single sub-field is very difficult. Setting priorities across fields of research, as the AASC was required to do, is even harder. To help with this daunting task, the committee followed precedent and divided all proposed initiatives into three categories—major, moderate, and small—based on cost. Initially, ground-based initiatives were separated from space-based ones as well. Because space-based projects are generally much more expensive than ground-based ones, the cost categories are different for the two cases: for ground-based initiatives, moderate projects are those with a decade cost of $5M to $50M, whereas space-based moderate projects have a decade cost between the Explorer mission cap of $140M and $500M. The priorities recommended by the AASC are shown in Table 1, which is taken from *Astronomy and Astrophysics in the New Millennium* (NRC 2001).

The top priority among major initiatives for the coming decade is the Next Generation Space Telescope (NGST). This is a large (~6.5m), passively cooled telescope in space that will be approximately 100 times as sensitive as the Hubble Space Telescope (HST) or the Space Infrared Telescope Facility (SIRTF) in the infrared. NGST is designed to reveal the first epoch of star formation and to follow the evolution of galaxies from their birth to the present. With an angular resolution several times better than HST, it will provide a unique capability for studying the formation of stars and planets in our own Galaxy as well.

Table 1. Prioritized Initiatives (Combined Ground and Space) and Estimated Federal Costs for the Decade 2000-2010[1,2]

Initiative	Cost[3] ($M)
Major Initiatives	
Next Generation Space Telescope (NGST)[4]	1000
Giant Segmented Mirror Telescope (GSMT)[4]	350
Constellation-X Observatory	800
Expanded Very Large Array (EVLA)[4]	140
Large-aperture Synoptic Survey Telescope (LSST)	170
Terrestrial Planet Finder (TPF)[5]	200
Single Aperture Far InfraRed (SAFIR) Observatory[5]	100
Subtotal for major initiatives	2,760
Moderate Initiatives	
Telescope System Instrumentation Program (TSIP)	50
Gamma-ray Large Area Space Telescope (GLAST)[4]	300
Laser Interferometer Space Antenna (LISA)[4]	250
Advanced Solar Telescope (AST)[4]	60
Square Kilometer Array (SKA) technology development	22
Solar Dynamics Observatory (SDO)	300
Combined Array for Research in Millimeter-wave Astronomy (CARMA)[4]	11
Energetic X-ray Imaging Survey Telescope (EXIST)	150
Very Energetic Radiation Imaging Telescope Array System (VERITAS)	35
Advanced Radio Interferometry between Space and Earth (ARISE)	350
Frequency Agile Solar Radio telescope (FASR)	26
South Pole Submillimeter-Wave Telescope (SPST)	50
Subtotal for Moderate Initiatives	1,604
Small Initiatives	
National Virtual Observatory (NVO)	60
Other Small Initiatives	246
Subtotal for Small Initiatives	306
DECADE TOTAL	**4,670**

[1] Cost estimates for ground-based capital projects include technology development plus funds for operations, new instrumentation, and facility grants for five years.

[2] Cost estimates for space-based projects exclude technology development.

[3] Best available estimated costs to U.S. Government agencies in millions of FY 2000 dollars and rounded.

[4] Cost estimate assumes significant additional funding to be provided by international or private partner.

[5] These missions could start at the turn of the decade; only estimated costs for this decade are included.

The highest priority for a major ground-based initiative, and the second priority overall, is the Giant Segmented Mirror Telescope (GSMT). This 30-m class telescope will operate in the atmospheric windows between 0.3 and 25 μm. Adaptive optics will give it diffraction-limited performance down to about 1 μm. Just as the Keck Telescopes have complemented HST by obtaining spectra of the objects imaged by the smaller space-based telescope, so GSMT will complement NGST. The powerful spectroscopic capability of GSMT will allow it to study both the evolution of galactic structure and the details of protoplanetary disks. Substantial progress in adaptive optics and in the design of large telescopes is needed before GSMT can become a reality.

The highest priority among the moderate initiatives is the Telescope System Instrumentation Program (TSIP). Universities and private observatories operate most of the optical and infrared telescopes in the U.S., yet many of these telescopes have inadequate instrumentation. Under TSIP, the NSF would fund the construction of peer-reviewed instrumentation at private observatories in exchange for telescope time or equally valuable benefits for the community. TSIP will have a multiplier effect by improving the capability of telescopes that have already been constructed. While it is likely that most of the approved instruments would be on large telescopes, instrumentation on small telescopes is by no means excluded. By expanding the number of instruments being designed and constructed at universities, TSIP will provide an important boost to the education of the next generation of instrumentalists.

The second priority among the moderate initiatives is the Gamma-ray Large Area Space Telescope (GLAST), which will dramatically improve our ability to study sources of energetic photons between 10 MeV and 300 GeV. The remaining priorities in this category are the Laser Interferometer Space Antenna (LISA), which will be able to detect gravity waves from merging supermassive black holes throughout the visible universe and from close binary stars throughout the Galaxy; the Advanced Solar Telescope (AST), a 4-m class telescope with a powerful adaptive optics system that will revolutionize the study of magnetic fields on the surface of the Sun; technology development for the Square Kilometer Array (SKA), an extremely powerful centimeter-wave telescope to be built a decade from now; the Solar Dynamics Observatory (SDO), which will study the outer convective zone of the Sun and the structure of the solar corona; the Combined Array for Research in Millimeter-wave Astronomy (CARMA), a powerful millimeter array in the Northern hemisphere that will study star formation at all epochs; the Energetic X-ray Imaging Survey Telescope (EXIST), which will survey the sky every 90 minutes from 5 to 600 keV, searching for weak, time-variable X-ray sources; the Very Energetic Radiation Imaging Telescope Array System (VERITAS), which will complement GLAST by studying photons between 100 and 10,000 GeV; Advanced Radio Interferometry between Space and Earth (ARISE), which will improve the angular resolution of very long baseline interferometry by a factor of 6; the Frequency Agile Solar Radio telescope (FASR), which will study the Sun over the frequency range 0.3-30 GHz; and the South Pole Submillimeter-wave Telescope (SPST), which will survey the dusty universe, identify primordial galaxies, and study small variations in the cosmic microwave background radiation. Many of these moderate initiatives involve small telescopes.

The top priority among the small initiatives is the National Virtual Observatory (NVO). By combining all major astronomical data archives into a single digital database and providing the techniques and services required for mining this vast database, NVO has the potential to enable qualitatively new discoveries. NVO will permit real-time observations to be accessible over the Internet, making it a powerful resource for public education and outreach. Small telescopes, such as those used for the 2MASS survey and the Sloan Digital Sky Survey, will make a critical contribution to NVO.

The committee recommended a number of other small initiatives that were not prioritized. Of particular importance for small telescopes, the committee recommended that NASA invest in Ultra-Long Duration Balloons, which offer the prospect for carrying instruments weighing up to several tons for periods of up to several months. The committee also endorsed the continuation of a vigorous Explorer Program by NASA.

Small telescopes are of particular importance in the training of the next generation of astronomers. The committee recommended that universities "should assume the responsibility for purchasing, instrumenting, and operating *small* (emphasis in original) telescopes needed by their students and faculty," just as they have long provided the laboratories needed for the education and research of biologists, chemists, and physicists.

When we look back on the first decade of the 21st century ten years from now, we can hope that there will have been a rich harvest of discoveries. As in the past decade, these discoveries will have been made with telescopes of all sizes, and we can expect that the students of that time will continue to receive essential training on small telescopes.

REFERENCE

National Research Council 2001, *Astronomy and Astrophysics in the New Millennium*. National Academy Press, Washington D.C.

Chapter 3

Community Perceptions of the Relative Worth of Large and Small Telescopes

Leslie J Sage
Nature
Washington, DC USA

Abstract: It is hard to think of an issue in astronomy more likely to generate controversy and bad feelings than the comparison of the scientific value of small and large telescopes. This is not just a matter of money, papers cited, and 'impact', for if it were the extensive literature on publishing trends would already have led to some general and widely accepted conclusions. Moreover, we would now have a coherent strategy for the future benefit of all astronomers and our science.

Although the debate usually rages around money and citations, the underlying causes of the intensity instead relate to very personal and emotional issues such as jobs, self worth and individual psychology. By psychology, I mean that some people are naturally inclined to dig very deeply into particular problems, with the goal of finally resolving them. Other people prefer to get a zeroth or first order answer to what the bigger picture looks like. Both approaches are essential to the success and vitality of any science. Coupled with these issues are individual preferences for sticking with instruments you know, or wanting to learn something completely new. Some astronomers are passionately committed to the concept of national facilities, and others are equally committed to their demise (or at least emasculation), sometimes under the assumption that money saved there would automatically go to their private observatories. These personality differences have become increasingly important because of some recent factors largely out of the control of astronomers.

We have gone through a relatively bad decade with fairly flat funding to the operations budgets of the national facilities and individual grants programs (though new money came into astronomy through new facilities such as Gemini and the Very Long Baseline Array), at a time when National Science Foundation (NSF) policies actively encouraged the overproduction of PhDs. Moreover, as we seek to build new telescopes, especially on new sites, there is increasing resistance amongst environmentalists, who often view us as being greedy in wanting all high mountains for ourselves. As environmental awareness increases in developing countries, we must be more vigilant and selective in choosing priorities.

T.D. Oswalt (ed.), The Future of Small Telescopes in the New Millennium, Vol. I, 13–19.

Changes in how we do our science, how we view ourselves as scientists, and how we interact with the wider world all are necessary. I do not see any quick fixes or easy choices ahead of us, but I hope that if we can acknowledge the underlying roots of our conflicts and face squarely our inadequacies in dealing with the public, we can make the difficult decisions that will benefit astronomy as a whole.

Keywords: small telescopes, productivity, science policy, public perception

1.   INTRODUCTION

It is very difficult to write an article about "community perceptions" of small telescopes, because there is no community consensus. Indeed, there is no single "community" from which to draw perceptions. Almost every telescope has its own community and its own niche in the wider picture. Instead of trying to represent a non-existent community view, I will simply discuss my own perceptions from my perspective as the astronomy editor of *Nature*. Some of these perceptions will not accurately reflect background events of which I have no knowledge, but on the other hand, there are many instances where background events simply are not communicated effectively to the community at large (or privately to me, even when I ask).

During my time at *Nature*, the Keck telescopes, the ESO Very Large Telescope (VLT) and Gemini North have opened, the Hubble Space Telescope (HST) has been repaired and BeppoSAX was launched. Adaptive optics systems have been built and installed on several large and medium-sized telescopes, where resolutions comparable to or better than that of the HST are now being achieved. These developments have had an almost immediate impact on astronomy. The Keck telescopes in particular have been extremely influential by accomplishing in one or two nights projects that were essentially impossible to do on 4-5m class telescopes.

2.   THE MYTH OF EFFICIENCY

A lot has been written about the "efficiency" and "cost effectiveness" of telescopes (Abt 1980, Trimble 1995, Leverington 1996), with several recent studies (Gopal-Krishna and Barve 1998, Leverington 1997a, Leverington 1997b, Benn et al. 2001, Hellemans 2000) trying to get to grips with the more difficult issue of relative worth. Although the studies use, broadly speaking, similar datasets or at least similar selection criteria, often the conclusions are at odds with each other: the most striking example of this is seen in Leverington 1996 and Gopal-Krishna & Barve 1998. Gopal-Krishna

& Barve (1998) conclude that in ground-based optical astronomy small telescopes (those with primary diameters < 2.5m) still play an influential role, while Leverington (1996) shows in his Figure 3 the declining influence of telescopes < 2.54m. The origins of such discrepancies can be difficult to determine, because often the actual data are not presented and subtle biases can creep into the analysis. In the case of Leverington (1996) and Gopal-Krishna & Barve (1998), much of the explanation is that Gopal-Krishna and Barve selected all astronomy papers in *Nature*, which typically include ~20 percent Solar System studies, while the sample in Leverington (1996) was drawn from the *ApJ* and *MNRAS*. There is little planetary science in the *ApJ*, and effectively zero in *Monthly Notices*, so the two samples are not really comparable. Both draw 'correct' conclusions from the data, but the conclusions differ because the initial conditions were different. It must also be said that the biases of the authors upon embarking on these studies help to determine the samples, and therefore largely predetermine the outcome.

I often find arguments about efficiencies and cost effectiveness to be misleading. Take for example the issue of productivity (papers produced), or productivity per dollar spent. Although the HST has produced many fine papers, its line in NASA's 2002 budget runs about $162 million (down substantially from its peak, and this does not include shuttle support costs!). Leverington (1997b) and Benn & Sanchez (2001) both conclude from this that the HST is not really cost-effective, though both state that the HST has been very influential in terms of science.

To me, a better way to look at the situation is to ask whether there is any possibility of doing the most influential science accomplished by (say) the HST from the ground. In some cases – particularly in the ultraviolet – the answer is no. Cost effectiveness is not an appropriate metric when you have a facility that can do what no other can. The one place where a discussion of efficiency or cost effectiveness makes good sense is in determining whether projects can successfully be accomplished using a smaller telescope, or ground-based rather than space-based, thereby freeing time on larger (or more expensive) ones for the critical observations that only the larger or space telescopes can accomplish. It may be useful to discuss whether the unique science done by the HST has been worth the cost, but it should be borne in mind that much of that money probably would not otherwise have been spent on astronomy.

3.   A SOLID NATIONAL INFRASTRUCTURE IS IMPORTANT

At the same time, it must be remembered that you cannot give a student hands-on training – which must include the ability to make mistakes – on the HST. Even the Very Large Array has a policy against use by inexperienced students. The case for retaining small telescopes therefore includes an important component of supporting the national science infrastructure. Moreover, as will be detailed in other chapters, there are numerous long-term monitoring projects, wide-field studies, etc. for which dedicated (and generally small or medium-sized) telescopes are best suited. There remain many problems with our understanding of the stellar contents of the Milky Way, and even in the last five or so years new dwarf galaxies have been discovered as hitherto unknown companions to the Milky Way or M31. It should be remembered too that the Palomar Sky Survey was done with an 18" telescope! The modern versions (Sloan Digital Sky Survey and 2MASS) are being done with telescopes whose sizes are modest in comparison to the Keck, VLT and Gemini telescopes.

However, it also is important to consider the total cost of running all the small telescopes and their relative contribution to the overall vigor of scientific research. Here, Leverington's (1996) analysis is useful, because it does show a trend towards the declining influence of research at small telescopes. (One can debate the zero point, but the overall trend is clear.) In a sense, this is to be expected as the 'easier' projects are done, with 'harder' ones requiring larger collecting areas. This is one of the reasons that the National Optical Astronomy Observatory (NOAO) has chosen to move towards 2-3m class telescopes as their new range of 'small telescopes', and for a national facility this seems sensible and defensible.

Kitt Peak National Observatory has attempted to ease the transition, through the 'privatizing' of smaller telescopes (and absorbing some budget cuts themselves for a year or so before closing telescopes). A good example of the former is the 0.9m SARA telescope. A consortium of southeastern universities runs it for about $50,000 a year, which includes a 'footprint' lease to NOAO. The process of privatization can take years, however – in the interim, such telescopes are closed and any other interpretation is simply 'spin control'. The 1.3m telescope is an example of that.

4. PSYCHOLOGY – AND POLITICS – ARE IMPORTANT

The transition to an era in which the smallest national research telescopes are of order 2m (or even larger) will inevitably cause debate and trouble as people lose their favored and favorite instruments. In general, the main stated worry is whether ongoing research programs can be maintained and finished in an orderly way, but the underlying concern really is that some researchers feel that they will be left unable to compete successfully for telescope time at the new, larger facilities. Without telescope time, they will not be able to write papers, attract students, and in some cases may not even be able to keep their jobs. Unless these very real worries are addressed in a careful and sensitive way, the closure of telescopes will inevitably create pointless controversy.

What should happen is that long-range planning should specifically include the phasing out of facilities as new ones open – and such matters should be discussed even when the new facility is little more than a gleam in someone's eye — so that a community consensus about timescales and priorities can evolve.

A particularly interesting case study in how not to manage a telescope closure occurred within the radio community in the spring of 2000, when the National Radio Astronomy Observatory (NRAO) announced without any warning that the 12m telescope on Kitt Peak (a facility with which I am well acquainted) would close at the end of the observing season (mid July). There was no consultation at all, and the community was greatly surprised because the expectation had been that the 12-m telescope would stay in operation until the Atacama Large Millimeter Array (ALMA) construction was underway. There were outraged protests to the NSF and to Congress, petitions to NRAO and stories both in the popular press (in Arizona) and the scientific press (*Science, Nature, etc*). Enormous bad feeling towards NRAO was generated, with a relatively trivial cost savings (only staff with direct operations responsibilities were let go – NRAO continues to have a very large Tucson office). Instead of continuing the strategic vision with which there was broad community acceptance – orderly closure of the 12-m at an appropriate time during the construction of ALMA – NRAO annoyed a lot of people with little substantive gain to itself. Moreover, the user community supported by the 12-m telescope is not adequately served by the other options available (a percentage of the times available on the OVRO and BIMA arrays, and the aging FCRAO telescope, which is in a poor location). The reason put forward for the closure was fiscal constraints, but this has never been detailed for the community. Private conversations with several key individuals suggest that the cost savings during the first year or two of closure are of order one percent of the NRAO budget, which seems like a bad bargain to the users concerned.

5.   CHALLENGING TIMES ALSO PRESENT OPPORTUNITIES

Probably unlike some other authors in this volume, I do not see this as a time of doom and gloom. A time of transition away from small nationally supported telescopes in North America and Europe should be regarded as an opportunity for broadening the base of astronomical research throughout the world. NOAO's policy of privatization of small telescopes could be broadened to include potential nationally funded facilities for other countries. Many developing countries now have substantial populations of scientists, for which some smaller telescopes on developed sites could provide the nuclei on which to build. Telescope facilities long since paid for by developed countries could be transferred at little or no cost to national funding agencies of developing countries, thereby keeping them open and productive, while allowing resources to be allocated to the new – and much more expensive – facilities. If the developing countries need additional financial support, there are options open to them (say, through the UN or the World Bank) that are not open to developed countries. If Western countries remotely operate telescopes in Chile, why should India or China not run telescopes on Kitt Peak, Mauna Kea or the Canaries? These telescopes could continue the admirable and successful strategy of encouraging applicants from the worldwide community – the competition would be good for everyone, and the telescopes would remain an active part of the international astronomical community.

Numerous environmentalists oppose the development of new observatory sites – Mount Graham is an example – and there remains substantial (and increasing) opposition to the presence of telescopes on Mauna Kea. Such concerns are becoming an issue in developing countries as well. In order for astronomers to be perceived as being sensitive to environmental concerns, we need to limit consciously our developments, and carefully assess priorities so that when we really want a new site we have some reasonable chance of getting it. Making use of existing observatory locations is one way in which we can show environmentalists that we do not want to take over all high, dry mountains.

My hope is that as plans are drawn up for the next generation of telescopes – the 30m to 100m monsters now being talked about – administrators of the present 3-5m class telescopes will start talking openly about plans for orderly transitions of their facilities. Some thought should be given even now to the times and conditions for reallocating resources away from the 6.5-10m class telescopes now in operation or under construction.

Change will inevitably hurt some people, but well-planned change can minimize the hurt so that we can present to politicians and the general public a true consensus that will leave the next generation of astronomers a legacy of which we can be proud.

6.   REFERENCES

Abt, H.A. 1980, PASP 92, 249.
Benn, C.R. & Sanchez, S.F. 2001, PASP 113, 385.
Gopal-Krishna & Barve, S. 1998, Bull. Astr. Soc. India 26, 417.
Hellemans, A. 2000, Nature 408, 12.
Leverington, D. 1996, Q.J.R. astr. Soc. 37, 643.
Leverington, D. 1997a, Nature 385, 196.
Leverington, D. 1997b, Nature 387, 12.
Trimble, V. 1995, PASP 107, 977.

Chapter 4

Funding of the Essential Synergy Between Small and Large Telescopes

Wm. Bruce Weaver
Monterey Institute for Research in Astronomy
Marina, California USA

Abstract: Since the 1956 Flower and Cook Symposium on the future of moderate-sized telescopes, such telescopes have contributed a substantial fraction of the important scientific results in astronomy. In addition, some of the significant science and discoveries produced by astronomers at large telescopes could have been made at smaller instruments at that time. When considered on a per dollar basis, the scientific productivity of smaller telescopes dominates the field. Some small telescope myths, such as "small telescopes are best dedicated to large survey projects" are examined. If "goodness" of science is defined in any terms other than its cost, smaller telescopes easily hold their own. Astronomy has a history of an essential synergy between small and large telescopes. This synergy can be maintained only if there are a reasonable number of well-maintained, well-instrumented smaller-sized telescopes. Recommendations for a small telescope budget and a refurbishment of the Schmidt telescopes at the national observatories are made. Two percent of the Astronomy and Astrophysics Survey Committee decadal budget would provide well-instrumented facilities to the same number of astronomers who use the existing or approved large ground-based optical and infrared telescopes and add 50% to the amount of quality astronomy produced.

Key words: research productivity, small telescopes, large telescopes, decadal survey

"In publishing the original Letter by H.L. Johnson and W.L. Richards (Ap.J.[Letters], 160, L111, 1970) the Editor did not sufficiently realize that the subject is a highly controversial one in which divergent views are strongly held. It does not appear that any useful purpose will be served by publishing further correspondence on this matter in the pages of the Journal.

T.D. Oswalt (ed.), The Future of Small Telescopes in the New Millennium, Vol. I, 21–39.
© 2003 *Kluwer Academic Publishers. Printed in the Netherlands.*

Accordingly, with the publication of Dr. Okes's Letter stating a different point of view the subject is closed"[1]

1. INTRODUCTION

The subject was the optimum size of optical telescopes and it was hardly closed; only banned from the pages of the Astrophysical Journal. The purpose of the current contribution is to provide some thoughts on an optimal expenditure of funds in astronomy with respect to the issue of telescope size. This purpose is narrower than the issues addressed by the Astronomy and Astrophysics Survey Committee (AASC, 2001, hereafter AASC) but integral to its general charge of how to allocate resources in the coming decade. In order to address this issue, we must consider the relative contribution to astronomy of the various components and their costs.

In 1956, the dawn of the post-photographic era and its effects on the relative importance of small telescopes to the advance of astronomy was heralded by a two-day symposium in honor of the opening of the Flower and Cook Observatory: *The Present and Future of the Telescope of Moderate Size* (Wood 1958). As Wm. Hiltner observed to open the meeting, "Most astronomers are faced with the problem of having a telescope smaller than the largest." Although CCDs were not yet on the horizon, techniques by Lallemand, Hiltner, and Fellgett in electronic image detection led to the observation that modest-sized telescopes using such detectors could perform as well as or better than the gold standard 200-inch telescope in limiting magnitudes while using photographic plates.

The promise of that meeting has now, nearly 50 years later, been fully realized as back-illuminated CCDs with nearly perfect quantum efficiency and almost no read-out noise are within financial reach of almost all astronomers. Such detectors make their modest-sized telescopes equivalent to one ten times their size in the photographic era, usually exceeding the performance of a 5-meter telescope[2]. Of course, large telescopes have comparable or better

---

[1] S. Chandrasekhar (1970).

[2] E.g., Deeg and Ninkov (1996) performed V, R, and I photometry to 22.5 magnitude with a 36-inch telescope. Better detectors on the same telescope have now lowered the threshold fainter than 23rd magnitude. This is comparable to photographic results reported by Baum (1962) for the 5-meter telescope. Fry et al. (1999) reached 27.5 R magnitude arcsec[-2] with a 24-inch aperture. Low-resolution spectroscopy of 19th and 20th magnitude objects is not unusual with 1 – 2 meter telescopes (e.g., Meusinger & Brunzendorf 2001). This compares favorably to a

detectors but, in this 50-year grace period, the 5-meter telescope certainly did not exhaust the important astrophysics then available only to it.

The current trend, in most countries, is to close smaller telescopes in order to support fewer, larger telescopes. At the American national observatories, this represents an important shift from their original charge of providing telescope access to many astronomers from institutions without reasonable observing facilities and building unique equipment for all the astronomers of the country[3]. The recommendations of the AASC support the continuation of that trend. Even with the large telescopes absorbing well over 90% of the astronomy funding in the U.S., their proponents have appealed to users of smaller telescopes to support an increase of that fraction![4]

2. SCIENCE EFFECTIVENESS OF SMALL TELESCOPES

The difficult point to consider is what is the relative science effectiveness of small telescopes. Obviously, for observations at the limits of practical observational patience, large telescopes are essential. Also, unless the telescope has unique properties other than its size, it is difficult to justify instrumentation that is orders of magnitude more expensive than the telescope.

First, it seems worth stating that the *cost of a scientific result is not a measure of its scientific value*[5]. In astronomy, where serendipity is common, it is probably more difficult than for most sciences for expensive projects to rise above the mediocre because, understandably, funding agencies are reluctant to risk large sums of money on projects with low or uncertain probability of success.

Are citations a reasonable measure of scientific importance?

Counting citations has been criticized as unfairly favoring smaller telescopes because they are more likely to be involved in survey projects which, in amassing large amounts of data, are likely to be frequently cited. It is not clear that this bias exists. Certainly, the greater pressure on larger instruments means that projects with shorter telescope runs are favored. Thus, a wider variety of

---

limiting magnitude of 17 reported for comparable dispersions with the 5-meter (e.g., Greestein and Sargent 1974)

[3] Supporting 700 different astronomers and students and producing 350 papers annually (Abt, 1985)

[4] Lowell Observatory Fall Workshop: The Role of Small Telescopes in Modern Astronomy, 1996.

[5] My favorite version of this is due to an anonymous NASA referee who asked why, if the science in the proposal was as good as it seemed, it was not proposed for a larger telescope?

papers are produced per night on a larger telescope. This produces a higher paper count per night and provides a greater opportunity for high citation papers.

The argument that survey projects unfairly favor small telescopes is based on certain assumptions. First, that the large amount of effort that goes into a major survey is somehow less deserving of its citation count than some different form of effort that goes into other research projects. Second, that the final product of a large survey is of less scientific importance than other research.

Surveys are performed when little is known about what a new observational opportunity will reveal (e.g., first X-ray satellite) or when the scientific question requires statistically complete data (e.g., Galactic structure, distribution of mass in the Universe). In the latter case, this usually reflects a maturing of the discipline; enough of the science involved is understood well enough that the extensive use of resources is warranted to produce the desired results.

On a scientific basis, the decision to commit resources to a major survey seems to be more dependent on the state of the scientific understanding than the size of the required instrument[6]. Although survey work has been frequently given as a possible justification for small telescopes, surveys are now becoming more prevalent on large and expensive telescopes as well. For example, the DEEP (The Hubble Deep Fields and Deep Extragalactic Evolutionary Probe) survey proposes to use nearly 200 Keck nights.

Surveys are the prime example of a distinction made by Abt (1996, 1999) between papers that are heavily cited because of their fundamental nature and those that are heavily cited because of their usefulness. Such a distinction is important in selecting papers for a Centennial volume (Abt 1999), but useful papers, which are often papers tabulating the results of surveys, are certainly as important to our ultimate understanding of how things work as those that we acclaim as fundamental. Abt (1996) lists the 15 papers from 1954 that were the most cited in the following 40 years. Roughly half were in the fundamental category and half were in the useful category.

Another difficulty with using citations as the measure of scientific value is that while, in the full course of history, a set of scientific results or ideas may be judged of great scientific importance, its citation rates in its time may be low because of slow recognition of its significance in the community[7], a lack of

---

[6] On a cost and political basis, it is obviously more difficult to commit a very expensive instrument to long projects.

[7] E.g. Practical microlensing (Paczynski 1986) or Zwicky (1957); although one of the most famous cases of the 20th Century is probably that of biologist Barbara McClintock.

current popularity in that specialization, or an inability of the current technology to address the ideas[8].

Alternative methods of ascribing scientific worth might include the practical importance to mankind. In that case, the search for Earth-crossing asteroids would have to take first priority and SETI projects would probably be second[9].

It is so difficult to provide a quantitative method, outside of citation counts, to assess the scientific value of any research that one must consider *ab initio* to assess all valid scientific inquiry of equal value. This may be a more reasonable approach in astronomy than in many other sciences considering the substantial interdependence across specialties and the strong role of serendipity. For example, it is hard to imagine a field of astronomy that does not critically depend on positional astronomy as an essential building block to its results. This approach is difficult to quantify because of differing authors publishing styles and the nature of the results (e.g., survey versus discovery).

In short, citation count is certainly an imperfect method for determining scientific productivity but seems to be the most easily quantifiable and least controversial. However, using either numbers of papers or numbers of citations produces the same results: the majority of research is accomplished on small telescopes. Abt (this volume) finds those telescopes 2.5 meters in aperture or smaller produce 62% of the papers and are responsible for 55% of the citations. He finds that 75% of the citations are accorded to research performed with telescopes of less than 4-meter aperture.

3. COST EFFECTIVENESS OF SMALL TELESCOPES

Abt (1980) established the cost effectiveness of small telescopes, in terms of cost per citation. He found that the initial cost at KPNO for buildings, dome, mounting, and optics was equal to $362,000 $A^{2.37}$ for the 0.4 through the 4 meter telescopes where A is the aperture. For the years 1973-78, Abt found that the citation rate was proportional to $A^{1.5}$. Thus, roughly, *the dollar cost per citation is directly proportional to the aperture.*

A current estimate, based on slightly different parameters, provides a result even more favorable to smaller telescopes. Figure 1 shows the current cost (Melsheimer 2001) of only the telescope as a function of aperture of the

---

[8] E.g., Kuiper belt objects.

[9] Historically, although it may not have seemed practical at the time (or even now), the source of solar energy fits this category.

telescope. Note that the exponent is nearly the same as that found by Abt for both telescope and facilities but somewhat smaller than the 2.6 found for just telescope costs by Meinel in 1980 (Robinson 1980). The cost per collecting area increases with the aperture such that the cost per unit collecting area of a 10-meter telescope is nearly twice that of a 1-meter telescope. In this volume, Abt found that telescopes of apertures from five to ten meters produced only twice as many citations per paper as those based on data from telescopes of one to two meters in aperture. This suggests a cost per citation proportional to about $A^{1.5}$.

Telescope Costs

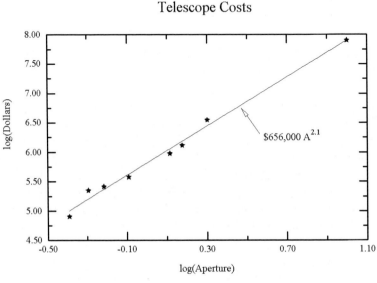

Figure 1. Current telescope construction costs (from Melsheimer, 2001.) Related costs, such as building or instruments, are not included.

Including the operating cost of the telescopes does little to change this latter relationship. Operating costs for one to two meter-class telescopes are of the order of $100,0000 while those of 8 to 10 meter-class telescope are about 50-100 times as much. This again suggests a cost per citation proportional to about $A^{1.5}$.

Tables A.1 and A.2 in the Appendix, are taken from Ringwald et al. (2001) who examined the productivity of 22 ground-based optical telescopes for papers published in 1995 and cited in 1998. After the HST, the typical size for the 22 most productive telescopes in terms of citations/area is about 0.9 meters. Since cost per unit area rises only slowly with aperture, we can take the citations/area to be a good approximation of citations/dollar. The average citations/area of the

22 most productive small ground-based telescopes is four times that of the large telescopes.

Thus, the financial considerations coincide with what most astronomers should feel intuitively: use the smallest telescope practical for the observation. This is possible only if such telescopes, properly instrumented, are available.

4. THE QUALITATIVE PRODUCTIVITY OF SMALLER TELESCOPES

The quantitative productivity has been covered by Abt (1980, and, more currently, this volume). These contributions cover both the publication rates by telescope size and the citation rates to those papers. Is there some additional qualitative aspect that is overlooked by this statistical approach?

Constructing a fully satisfactory list of the most important astronomical research since the publication of *The Present and Future of the Telescope of Moderate Size* is not a feasible task for this author. But, for the purposes of our current discussion, such a list should include not only those scientific works attributed to the smaller telescopes of their time but should also include those observations that could reasonably have been made by such telescopes.[10]   For example, the discovery of the nature of 3C 273, was made with a 400 A/mm resolution spectrum obtained with the 200-inch telescope; however, a spectrum of that resolution of a 12.8 magnitude object was clearly within the range of modest-sized telescopes of that vintage.

One can, however, examine lists made by others and determine the size of the instruments used by their authors.

The AAS Centennial Committee selected 53 papers for republication in the *Centennial* issue of the American Astronomical Society (Abt 1999). I assigned

---

[10] A clear example of such a case is the identification of Sco X-1. Stephenson identified the object with objective prism plates taken in May 1965 through July 1966 with the Burrell 24/36-inch Schmidt telescope for the purpose of identifying the x-ray source. Johnson confirmed this identification with spectra from the KPNO 84-inch in June 1966 (Johnson & Stephenson 1966). Also in June and July, 1996, Sandage et. al (1966) observed the object with the 200-inch and the Tokyo Observatory 74-inch. Although the papers were submitted within 4 days of each other in August 1966, the discovery is generally attributed to the 200-inch (109 citations vs. 16 citations for the smaller telescope-based results).

papers primarily as supported by large or small[11] telescopes or strongly theoretical. If a primarily theoretical paper was based on data from an identifiably sized telescope, credit was given to that telescope size. Large telescope papers, the observations of which could have been done by small telescopes, were tabulated in both in the large telescope row and its own row.

Table 1. Number of Papers by Telescope Size in *Centennial*

Type	Number
Large Telescope	15
Small Telescope	16
Theoretical	19
Large, could have been small	2
Space-based	3

The AASC (page 18) provides a list of 12 accomplishments of the 1990s worthy of being included in a list of the important discoveries of the decade. The first was doubtlessly the most important as, if for no other reason, it answered a pressing astronomical question of over a century: the discovery of planets orbiting stars other that our sun. The first two telescopes that made these discoveries were certainly small and they continue to lead this research. Small telescopes also play significant roles in several others on their list, including microlensing, observations of the impact of Comet Shoemaker-Levy 9, and the optical identification of gamma-ray bursters.

While I can't create a personal list that is complete in any way, I would like to add a few of my favorite small telescope projects to those listed elsewhere in this volume. The resolution of the nature of Herbig-Haro objects depended on a telescope with a large enough scale to capture the wide nature of these jets. The discovery of both radio and optical pulsars are the work of small telescopes. The detection of 23 planetary or brown dwarf objects with a 1.2-meter telescope that produces radial velocities with 7-8 m/s accuracy (Naef et al. 2001).

There are numerous lists of potential research suited to small telescopes. These lists are either soon made obsolete by increased capabilities of small telescopes by improved instrumentation technology (e.g., Percy 1980) or attempt to avoid the problem by being impracticably vague. The many contributions presented at the Lowell Observatory Workshop on Small Telescopes, IAU

---

[11] This required a floating definition with time. For example, the Mt. Wilson 60-inch was taken as a large telescope until the institution of the 200-inch telescope; all the Mt. Wilson solar telescopes were taken as large; and all Schmidts were taken as small.

Colloquium 183, and this volume describing either current or very near future research are probably the most reliable guides to such topics.

In some cases, small telescopes have specific advantages over large telescopes. For example, spectroscopy-resolution spectrophotometry (e.g., Torres-Dodgen & Weaver 1993) requires that all the light from the object pass through the spectrophotometer aperture. This becomes increasingly difficult with the increasingly longer focal lengths of larger telescopes. Appendix 2 shows that, for telescopes larger than about 1 to 2 meters, optimum spectrophotometer design becomes problematic.

5. THE INSTRUMENTATION AND MAINTENANCE OF SMALL TELESCOPES

A telescope lacking good instrumentation is like a person without all their senses — NOAO Director Jeremy Mould (2001)

Any small telescope without reasonable instrumentation is of little or no use to astronomers. An assured way to reduce demand for a telescope is to limit its capabilities with inadequate instrumentation. There are many examples of this at national and private observatories.

Small telescopes can be well instrumented at a cost in proportion to the telescope cost. The Panel on Ground-Based Optical and Infrared Astronomy (GBOI Panel, 1995) proposed, for the National Research Council, that a new facility-class instrument be built for existing telescopes every five years at a cost of about 25% the cost of the telescope.

As emphasized in the AASC, telescope facilities must be maintained. This is as true for small telescopes as for large. What we have seen above is that these costs scale in a similar manner to telescope costs and, hence, provide the same cost-effectiveness in terms of science produced. Gemini operations, facilities, and maintenance costs are 10% annually of their construction costs. This is similar to costs for telescopes in the 1 to 2.5 meter class.

These two factors, quality instrumentation and reasonable maintenance, are probably the most significant factors limiting the scientific productivity of small telescopes. Limited instrumentation and telescope controls reduce the capabilities, effectiveness, and attractiveness of the telescopes. Astronomers often maintain smaller telescopes. This is not a job for which they are necessarily well suited; so, not only does it reduce the productivity of the telescope by preoccupying their astronomers with engineering rather than astronomical research, but the maintenance is often not accomplished well either.

The fact that smaller telescopes are less well maintained and instrumented raises an interesting question about their productivity. If they were instrumented and maintained at a level comparable to that of large telescopes, but at the lower cost proportional to their size, how much would their already impressive productivity increase? *What would keep them from becoming as scientifically productive as are their larger brethren, only with an emphasis on different research topics?*

Perhaps nothing. The KPNO 2.1-meter telescope is an excellent example of a small telescope that has received large telescope support. Ringwald et al. (2001) examined the productivity of 22 ground-based optical telescopes, including 13 in the 3-meter or larger class, for papers published in 1995 and cited in 1998. They showed that the 2.1-meter was 4th in total paper production, 6th in citations, and higher in citations per surface area than any telescope larger than it.

Trimble (1995) examined the productivity of telescopes larger than 2 meters for an 18-month period in the early 1990s and their citations in 1993. In general, her results concerning the relative productivity of private and national observatories confirm those of Abt (1985). She shows that the 2.1-meter telescope (there were no smaller telescopes in the 16 that she considered) was among the top three telescopes in terms of papers and pages of research produced and seventh in citations.

6.   THE SELLING OF NEW TELESCOPES

The motivation for larger or unique telescopes is easier to express than that for more modest instruments. These instruments *will* be able to make observations difficult or impossible for their predecessors and, while their greatest scientific value may come from discoveries impossible to predict, one can, with a fair amount of confidence, make predictions about what new science can be addressed if the instrument is constructed.

Proposers of a non-unique telescope of an aperture that will not exceed the world's largest telescope have a more difficult case to make in a national forum. They can only assert that the telescope will enable the research of a comparable number of scientists as its larger cousin, at a fraction of the cost per scientist, and its greatest scientific value will probably come from discoveries impossible to predict. But they cannot claim to be able to observe what no other instrument could observe[12].

This approach to the justification of the astronomical budget has become strikingly reminiscent of that of the particle physics community and, if continued, will probably suffer the same fate.

7. AN EXAMPLE CASE FOR A SPECIALIZED SMALL TELESCOPES: SCHMIDT TELESCOPES

Many of the large instruments recommended by the AASC are specialized telescopes. An example of a general purpose specialized small telescope is the Schmidt telescope. With the Michigan Curtis 24/36-inch Schmidt at Cerro Tololo now closed and its twin Warner and Swasey Burrell Schmidt at KPNO and the Palomar 48-inch Schmidt out of general use, *there are no significant Schmidt telescopes available to American astronomers.*

How important are Schmidt telescopes to modern astronomy? Using the ADS citation lists, I counted the number of citations to refereed papers published in 1994 and 1995 using Schmidt telescopes. These lists are incomplete so these results should be viewed as lower limits. Figure 2 shows the distribution of the citations to these 90 articles. Abt (1981) showed that the citation rate peaked about 5 to 7 years after publication and that papers averaged one citation per year over their lifetimes. I chose the 1994-95 interval to be as close as possible to current date to accurately as possible represent the effects of Schmidts with modern instrumentation while giving their citation rates time to mature. In fact, many of the articles still referred to data gathered with photographic emulsions. The citation rate is twice that of the average paper from Abt's study.

The Ringwald results in the Appendix show that the citation rate/(area or cost) of the Burrell Schmidt to be over four times that of the large telescopes. Three Schmidt telescopes are in the top 22 effective ground-based telescopes and their average citation rate/(area or cost) is about three times that of the large telescopes.

The 24/36-inch Schmidts, located at two excellent sites in both hemispheres and well-equipped with a range of objective prisms can perform absolute

---

[12] This situation might seem clearer if applied to a different type of scientific instrument, say, electron microscopes. Why would one build any microscope that would not exceed the resolving power of the world's best?

Schmidt Telescope Citations for 1994, 1995 Papers

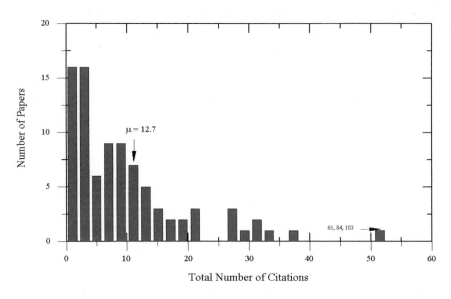

Figure 2. Total number of citations listed in ADS to refereed papers in 1994 and 1995 using Schmidt telescopes.

spectrophotometry over a 25 square degree-field to limiting magnitudes of 20. A large CCD array could cover a large fraction of that field. Accurate, automatic two-dimensional stellar classification by artificial neural networks (e.g., Bailer-Jones et al., 1998; Weaver & Torres-Dodgen, 1997) will be an essential component to the exciting next generation of Galactic structure and dynamics research enabled by deep-reaching satellite-based parallaxes and proper motions. Such instrumentation would also permit more complete, much deeper surveys of star formation regions or extragalactic emission-line objects (e.g., the KISS project, Salzer et al. 2000). Surface photometry of galaxy disks by Fry et al. (1999) reached 27.5 R magnitude arcsec^{-2} with the Burrell Schmidt.

Other Schmidts have been very productive lately with deep QSO surveys (e.g., Croom et al., 2001; Meusinger & Brunzendorf, 2001), detecting young objects near the Galactic center (e.g., Dufton, et al., 2001), searches for very low mass stars and brown dwarfs (e.g., Barrado y Navascués, et al., 2001), and Trans-Neptune Objects (Ferrin, I. et al., 2001).

The 24/36-inch Schmidts were not as oversubscribed while they were available at the national observatories as were the 4-meter telescopes, although both typically were oversubscribed in the late 1990s by 20 to 30%, ranging as high as an oversubscription factor of 2.9. This lower subscription rate may be because a typical night at a Schmidt telescope can generate enormous amounts of data (hundreds of spectra per hour and ten to a hundred times as much photometry) and because the telescopes have been poorly instrumented compared to their full capacity and compared to the large telescopes.

Currently, these telescopes sample only a small fraction of their 5-degree fields with single CCDs and are not generally available to American astronomers because of lack of national funding. These telescopes should be equipped with large CCD arrays and, as funded through the national observatories, be made available to general use.

Paving the focal planes of these telescopes with modern CCDs is an expensive project compared to the original costs of these telescopes but extremely inexpensive compared to the cost of projects listed in the AASC for the amount of astronomy that they could produce. The GBOI Panel has already proposed this for the Burrell Schmidt.

8. THE SUPPORTING ROLE OF LARGE TELESCOPES

The relationship between large and small telescopes is synergistic, each providing support for the other. The role of small telescopes in support of larger telescopes is often assumed and discussed; but the role of large telescopes in support of small ones needs further recognition.

There are at least a couple of natural ways in which this occurs. Often a research project can be largely accomplished with smaller telescopes but require the photon gathering power of a large telescope for project completion or a critical part of the data. An example of the latter, deconvolution of monitored images of gravitational lenses is substantially improved with HST images providing accurate size and position of the components.

A second way in which this supporting role is apparent reflects the continuing nature of astronomy as an exploratory science, still filled with frequent surprise discoveries. Astronomy is a science where observational results have a longer citation half-life (35 years) than theoretical papers (22 years, Abt 1996). As has been noted above, many discoveries are made with small telescopes where research with lower probability of success can be economically pursued. Once the discovery is made and the technique is vetted, the research, continued on small telescopes, is often supplemented or extended by large telescopes. The prime example must be the discovery of extra-solar planets and brown dwarfs but many others, such as gravitational lenses, rotation curves of galaxies, or Herbig-Haro jets are available as examples.

Conversely, discoveries made at large telescopes can often be most effectively exploited on smaller telescopes.

Small telescopes cannot uphold their half of the synergistic relationship with large telescopes unless they exist and are well-instrumented and well-maintained.

9.   CONCLUSIONS

1. By any measure, small telescopes produce the bulk of quality astronomical research.
2. Astronomy produced on small telescopes is much more cost effective than that produced on large telescopes.
3. The funding required to maintain the impact of small telescopes on the progress of astronomical research is a tiny fraction of the overall proposed budget for the current decade.

Astronomers do astronomy. Although many observational research projects can be accomplished in less time at larger telescopes, the extreme expense of the world's largest telescopes, which is their intrinsic nature, strongly limits the observing time available to much less than could be profitably used by all but a small fraction of the world's astronomers. With any balanced approach to telescope size distributions the cost of telescope construction and maintenance will be dominated by the larger instruments; however, it seems apparent that the amount of astronomy accomplished will be maximized if the cost of users of lesser-sized telescopes is a significant fraction of the total cost of astronomers.

Both quantitative and qualitative arguments demonstrate the continuing importance of small telescopes to the astronomical endeavor. The quantitative arguments show that it is significantly less expensive per citation to use the smallest telescope that will accomplish the research. Both the quantitative and qualitative arguments show that the research accomplished by small telescopes is of continuing and lasting significance to the discipline as witnessed by their non-diminishing contribution to astronomy over the last century and the persistence of their citation histories.

To continue the recent trend of closing many smaller telescopes to fund a decreasing number of very large telescopes assures a diminishing scientific productivity in astronomy. There has been a long history of synergy between large and small telescopes and breaking this highly productive tradition for a small fraction of the decadal funding proposed in the AASC would be very unfortunate.

A well-instrumented 1.5 meter telescope could be installed at an established site for about $2.5 M. A decade of maintenance is of the same order. *Two percent of the AASC $4,670M decadal budget would fund construction and maintenance of 17 such telescopes*, a number equal to the number of existing and approved large ground-based optical and infrared telescopes in the world (AASC, Table 3.3). This would double the number of astronomers with access to good telescopes and add something of the order of 50% to the amount of quality research (citations).

A more practical approach is suggested in Table 2.

Table 2. A Modest Decadal Budget for Small Telescopes

Number	Item	Cost	Total Cost
6	Instrumented 1.5 m telescope	$2.5M	$15M
3	Instrumented 2.5 m telescope	$6M	$18M
	Maintain above for decade	= construction cost	$33M
2	Schmidt Focal Plane CCD arrays	$2M	$4M
	Upgrade & maint. Existing telescopes		$30M
		Total	**$100M**

The suggested instrumentation for the proposed nine telescopes might be a relatively standard spectrograph and direct imaging CCD for each plus one relatively unique instrument for each of the telescopes.

The upgrading of existing small private and public telescope control systems and instruments is included with some maintenance support for these existing telescopes. This is similar to the $50M proposed by the AASC in their Telescope System Instrumentation Program, aimed at the upgrading of instrumentation of *large*, private telescopes. As the AASC states, such funding should be accompanied by maintenance funding. This is especially true for smaller telescopes, which are much more likely to be under-maintained.

If well instrumented and maintained, I'm sure that telescopes constructed for the national observatories or constructed at private observatories and made generally available would be well subscribed. They would be further over subscribed if TACs aggressively allocated observations to the smallest practical telescopes.

10. FINAL THOUGHTS

The projects proposed in AASC are meritorious; I'm sure most astronomers would rearrange a priority or two and/or substitute a program or two that didn't make the final list for one that did. It is difficult to believe, however, that either the stated individual budgets or the significance of the final decadal budget is accurate to the two percent discussed above. The substantial additional scientific productivity that would result from a small-telescope budget similar to the one proposed here or, conversely, the substantial loss of astronomical productivity if the proposed AASC zero budget for small telescopes were adhered to, is much more certain.

11. APPENDIX 1. PRODUCTIVE TELESCOPES

Table A-1. Top telescopes in terms of productivity per collecting area, solar and radio telescopes are omitted. From Ringwald et al., (2001, Table 7) who examined the productivity of 22 ground-based optical telescopes for papers published in 1995 and cited in 1998. Since cost per unit area of telescopes rises slowly with aperture, the citations/area is also a good representation of citations/dollar.

Table A-1. Top Telescopes in Productivity per Collecting Area

Telescope	Aperture (m)	Papers/area	Citations/area
IUE*	0.45	106	312
Mt. Hopkins APT	0.25	51	41
HST*	2.4	28	105
CTIO	0.9	18	37
Burrell Schmidt	0.6/0.9	18	28
KPNO	0.9	15	50
KAO	0.9	14	39
CBA West	0.36	13	13
CTIO Lowell	0.6	13	26
CBA	0.32	12	50
Goethe-Link	0.4	12	20
KPNO Coudé feed	0.9	12	27
U. Missouri	0.35	10	21
Las Campanas Swope	1.0	10	32
CTIO Yale	1.0	10	14
Mt. Laguna	1.0	9.4	12
CTIO Curtis Schmidt	0.6/0.9	9.2	10
U. Hawaii	0.6	9.2	11
Palomar	1.5	8.9	24
KPNO	1.3	7.7	20
KPNO	2.1	7.1	18
Landis	0.2	7.0	10
CTIO	1.5	6.9	18
Palomar Schmidt	0.45	6.3	13
Average of ground-based telescopes			**24.8**

* These two space-based telescopes are included for completeness and interest but are not included in the average. The IUE is somewhat outside the scope of this article but it certainly ranks as a small telescope/inexpensive science satellite that has produced enormous amounts of quality science.

Table A-2. Productivity for Other Telescopes (from Ringwald).

Telescope	Aperture (m)	Papers/area	Citations/area
CTIO Blanco	4	2.6	8.5
KPNO Mayall	4	2.5	11
CFHT	3.6	2.7	10
Palomar	5	1.2	3.9
MMT	4.5	1.4	5.2
Keck I & II	10	0.3	1.3
Steward	2.3	4.3	12
Lick Shane	3	2.5	7.4
IRTF	3	2.2	6.1
AAT	3.9	1.2	4.2
WHT	4.2	0.9	2.2
UKIRT	3.8	1.0	5.0
		average	6.4

12.   APPENDIX 2. SPECTROPHOTOMETER EFFICIENCY

The scale at the focal plane of a grating spectrograph is $S_{tel} \cdot F_{tel}/F_{cam} = (D_{tel} F_{cam})^{-1}$, where S_{tel} is the scale of the telescope, F_{tel} is the f-ratio of the telescope, D_{tel} is the diameter of the telescope, and F_{cam} is the f-ratio of the spectrograph camera. Thus, to collect all the light from an object,

$$D_{tel} F_{cam} \leq W_d / \beta(\cos\theta_i / \cos\theta_d),$$

where W_d is the linear width of the spectrograph entrance aperture at the detector, β is the angular size of the area of the sky being imaged onto W_d, and $\cos\theta_i / \cos\theta_d$ is the ratio of the cosine of the angle of incidence to the cosine of the angle of diffraction of the dispersing element, which we take as unity. For good seeing and tracking, and allowing for the differential refraction of typical hour angles, we take β as about 5 arcseconds and W_d, ideally, as two 25μ pixels (= 50μ) so

$$D_{tel} F_{cam} \leq 2,$$

where the diameter of the telescope, D_{tel}, is expressed in meters. Thus, as the f-ratio of spectrograph cameras are rarely lower than 2, telescopes larger than 1 meter must use more pixels than that suggested by the sampling theorem or adopt other strategies to accomplish spectrophotometry.

13. REFERENCES

Abt, H. 1980, PASP, 92, 249.
Abt, H. 1985, PASP, 97, 1050.
Abt, H. 1996, PASP, 108, 1059.
Abt, H. ed. 1999 The Astrophysical Journal American Astronomical Society Centennial Issue (Chicago: University of Chicago Press).
Bailer-Jones, C. A. L., Irwin, M., von Hippel, T. 1998, MNRAS
298, 361.
Barrado y Navascués et al., 2001 Ap.J. Sup., 134, 103.
Baum,W.A. 1962 Stars and Stellar Systems. Volume 2. Astronomical Techniques (Chicago: University of Chicago Press).
Chandrasekhar, S. 1970, Ap.J. 162, L77.
Croom, S.M. 2001 MNRAS 322, L29
Deeg, H.J. & Ninkov, Z. 1996 A&A Supl. 119, 221.
Dufton, P.L., Smartt, S.J., Hambly, N.C. 2001 A&A, 373, 608.
Ferrin, I. et al., 2001, Ap.J. 548, L243.
Fry, A., Morrrison, H., Harding, P., Boroson, T. 1999, A.J. 118, 1209.
Greestein, J.L. & Sargent, A. 1974 Ap.J. Sup., 28, 157.
Johnson, H. & Stephenson, C.B. 1966, Ap.J. 146, 602.
Melsheimer, F. 2001, private communication.
Mould, J. 2001, quoted in NOAO Newsletter June, 13.
Meusinger, H. & Brunzendorf, J. 2001 A&A, 374, 878.
Naef, D. et al. 2001 A&A, 375, 205.
Paczynski, B. 1986, Ap.J. 304, 1.
Percy, J. 1980, JRASC, 74, 334.
Ringwald, F.A., et al. 2001, BAAS, 32, 1428.
Robinson, L.J. 1980, Sky & Tel., 59, 469.
Sandage, A.R. et al 1966, Ap.J. 146, 316.
Salzer, J.J. 2000, A.J., 120, 80.
Trimble, V. 1995, PASP, 107, 977.
Torres-Dodgen, A.V. & Weaver, W.B. 1993, PASP, 105, 693.
Weaver, Wm. B. & Torres-Dodgen, A.V. 1997, ApJ, 487, 847
Wood, F.B., ed. 1958, The Present and Future of the Telescope of Moderate Size (Philadelphia: University of Pennsylvania Press).
Zwicky, F. 1957, Morphological Astronomy (Berlin: Springer).

Chapter 5

Small Telescopes: A Reporter's Perspective

Robert Irion
Contributing correspondent, Science magazine
Santa Cruz, California, USA

Abstract: Journalism tends to ignore the value of small telescopes across the world. The truth is, however, that small telescopes are very capable of producing results that inspire the public and can still perform extremely useful tasks.

Key words: telescopes: efficiency, perceptions

Like it or not, we live in a Guinness Book of World Records world. The fastest runners, the richest CEOs, and the cutest septuplets all gain attention while we watch them and marvel. We're fascinated by the extremes of human achievement, and the media are happy to oblige with bite-size, "Hey, Marge!" chunks.

Something similar happens in astronomy. The most distant quasar and the brightest gamma-ray burst get the headlines—and every six months, there's a farther one or a brighter one still. Meanwhile, few reporters take time to explain the careful research on galaxy evolution or star formation that lets astronomers put those superlative events into context. It's not really the writers' fault. It's just that in our Guinness world, context is the first thing to go. There isn't enough time or enough space.

That's why most people think the coolest things in the universe are found by big, costly telescopes. They are today's record-holders, so they get the most ink and air time. The twin 10-meter Keck Telescopes in Hawaii are undeniably sexy, as are the 8.2-meter Very Large Telescope quadruplets in Chile and the other 8-meter-class telescopes conceived at the millennium. Then there's the 2.4-meter Hubble Space Telescope, pedestrian in size by ground-based standards but breathtakingly expensive. These facilities have publicity machines to match their technological glamour, especially Hubble. A steady flow of press releases and pretty pictures, color-enhanced for a

T.D. Oswalt (ed.), The Future of Small Telescopes in the New Millennium, Vol. I, 41–48.
© 2003 *Kluwer Academic Publishers. Printed in the Netherlands.*

maximum wow, keeps the pricey facilities in the public eye. Needless to say, we don't hear nearly as much about astronomy's army of small telescopes with apertures of 2 meters or less.

Of course, this era of big glass is not just a media creation. Leaders of the astronomical communities in the U.S., Europe, and Japan have poured resources into the quest to catch more photons. (Huge donations from philanthropists who love astronomy have helped a great deal.) Probing the depths of the universe is one of the field's frontiers, and these days, it takes clever machines and a lot of money to push a frontier. The optical wizards aren't stopping, either. Plans are brewing for 30-meter or even 100-meter segmented mirrors that astronomers, straining for adverbs, call "extremely large" or "overwhelmingly large." We can expect the public, for whom size matters, to crave news of such projects.

One can argue that media coverage tilts toward these behemoths. As a result, perceptions of the roles of big vs. small telescopes in astronomy are a bit skewed. The Guinness factor plays a role, since reporters tend to mention a telescope's size only if it's one of the titans. Apertures get ignored or buried in most stories about major discoveries driven by small telescopes—the first transit of an extrasolar planet, for instance, or the accelerating expansion of space. But for the most part, small telescopes aren't on the radar screens of writers. Just a handful of journalists cover the field in enough depth to recognize that astronomers need both kinds of tools, day in and day out.

When I set out to explore this dichotomy about two years ago, I encountered what the editor of this volume indelicately calls "aperture envy." I heard some grumbles about uncritical press reports on big glass. I also heard genuine concern about dwindling funds in the U.S. for nationally owned—and hence, nationally accessible—small telescopes. Michael Castelaz of East Tennessee State University lampooned this trend in a paper published in International Amateur-Professional Photoelectric Photometry Communications (Castelaz 1999; originally, this was a memorable anonymous poster at an American Astronomical Society meeting). Funding agencies would gradually build bigger and bolder ground-based telescopes by closing small ones to save money in an era of declining budgets, Castelaz wrote. By extrapolating this trend, he foresaw a single U.S. facility: a $1.2 billion, 42-meter "Ultimate Telescope," open to just 44 astronomers per year. "At least journals would be a bit lighter," he observed.

However, words of praise were more common during my interviews than jabs about perceived slights. Astronomers were eager to explain how small telescopes can thrive in the shadow of giant new observatories. And they were equally eager to proselytize to their colleagues, most of whom still fight for scraps of telescope time during each observing cycle at the major facilities. "Small telescopes that used to be thought of as outmoded or too small to contribute are still very capable of producing exciting results," said

Greg Henry of Tennessee State University. "There aren't enough people taking advantage of them."

Such comments steered me away from unbalanced public perceptions. Rather, I became fascinated by how some astronomers have revitalized small telescopes to do research that others can't do—important studies, worthy of public notice. It struck me as a Darwinian process. The survivors have adapted their instruments to the changing landscape of astronomy, finding niches and specializing to perform certain tasks extremely well. Large telescopes at the top of the budgetary food chain can pursue a wide range of studies and survive, but small ones had better find something unique to chase.

Many chapters in this book describe those evolutionary strategies in great detail. My own investigation was more cursory, but I still managed to talk to about two dozen practitioners of small-telescope science in a big-glass world. Here, I offer some of their thoughts as appetizers for the pages ahead.

If one mantra stood out above all others, it was this: To stay up-to-date, automate. Robotic telescopes zip among preprogrammed targets in the sky more quickly and make observations more consistently than people ever could. Not only that, but they decide whether to even open the dome and take data. Human attendants aren't needed. "Robotic telescopes are ideal for performing repetitive tasks that require many nights throughout the year," said Alex Filippenko of UC Berkeley. "It all boils down to money and efficiency. You can't run these small telescopes today if they're expensive, and operators are expensive. So let's automate them and get high-quality data at a fraction of the cost."

It's not easy to create a productive robot from scratch. Filippenko's own system, the 0.76-meter Katzman Automatic Imaging Telescope (KAIT), was a decade in the making. At the outset, the idea was not popular. But now, KAIT is setting records for finding nearby supernovae: more than 220 after five years of operation. KAIT replaced a small manual telescope at Lick Observatory. Every three to five clear nights, it records images of several thousand galaxies, one by one, slewing from one galaxy to the next in seconds. An analysis program then flags whether any new flares of light have appeared in the galaxies since the previous images were taken. Berkeley students check KAIT's identifications the next morning to make sure they aren't asteroids, electronic blips, or some other artifact.

Filippenko uses major observatories as well, but he views small, efficient telescopes as complementary. "High-energy physicists often criticize astronomers for never being able to let go of old, nearly defunct telescopes," he said. "But unlike in high-energy physics, small instruments still can perform very useful tasks."

Another automated system that gained renown was the Robotic Optical Transient Search Experiment (ROTSE) at Los Alamos National Laboratory in New Mexico. ROTSE used small telescopes indeed: four paparazzi-style telephoto lenses on a platform capable of swiveling rapidly to any part of the sky. Astronomers built ROTSE to pursue the optical flashes from gamma-ray bursts. A spectacular success came on 23 January 1999, when it triggered on an alert from the Compton Gamma-Ray Observatory. Within 22 seconds, ROTSE captured the light from an extraordinarily bright burst several billion light-years from Earth. A similar apparatus, the Livermore Optical Transient Imaging System (LOTIS), watches for flashes from the Lawrence Livermore National Laboratory in California with four 0.11-meter lenses. Both the ROTSE and LOTIS teams have upgraded to bigger lenses for more sensitivity to fainter objects (two 0.45-meter telescopes and one 0.6-meter telescope, respectively).

The nation's hotbed for robotic telescopes is Fairborn Observatory near Nogales, Arizona. Engineer Lou Boyd oversees a dozen telescopes ranging from 0.25 meters to 0.8 meters. Astronomers from South Carolina to Vienna receive data from the machines over the Internet. In particular, Greg Henry at Tennessee State has seen his career transformed by his eight telescopes at Fairborn. "I'm literally getting 50 to 100 years' worth of data now compared to what I would get in a single year working manually," he said. "And the cost of operation is a tiny fraction of what it was."

Henry's forte is examining long-term changes in the brightnesses of stars like our sun, which is vital for understanding how our star may evolve and influence Earth's climate. The precision of each robot's data from night to night lets Henry calculate fluctuations as small as one ten-thousandth of a magnitude, the most precise readings ever achieved. That specialty landed him on front pages in 1999 when one of his telescopes saw the star HD 209458 dim slightly when its Jupiter-size planet passed in front of it—the first extrasolar transit seen.

"I am tackling problems of intense scientific interest," Henry said. "I could never get these data with manual telescopes. It would never be precise enough." And, it goes without saying, no large facility would have the luxury to devote night after night to scanning for minuscule changes in a single object. Perhaps the best part, Henry said, is his observing schedule from his Tennessee office: "I program them and go home and get a good night's sleep."

Another emerging strength of small telescopes are globe-girdling networks that teams of astronomers have arranged to conduct uninterrupted studies. One such array is the Whole Earth Telescope (WET), directed by Steven Kawaler at Iowa State University. The network consists of a dozen or more observatories worldwide that collaborate twice a year for about two weeks at a time. The network includes 1-meter to 2-meter-class telescopes in countries such as China, Honduras, Lithuania, and Brazil, where astronomers otherwise might not get the chance to participate in international projects.

WET's main quarries are pulsating white dwarfs, with periods of minutes to hours. A single observatory could not follow repeated cycles, but WET's facilities—all operating under identical observing protocols—can keep an object in its sights for an arbitrarily long time. "The idea behind WET is the inverse of the British Empire: the sun never rises on it," said one of the network's coordinators, Donald Winget of the University of Texas, Austin. "I consider the network as a single instrument. We can keep an object in continuous view." An unbroken stream of observations allows the WET team to conduct asteroseismology on the distant dwarfs and gauge their internal structures. This, in turn, helps calibrate the dwarfs' ages and improve them as galactic chronometers. Another Earth-spanning array achieves even more detailed results for our favorite star, the sun. The network, called GONG (Global Oscillation Network Group), watches the sun's constant pulsations nearly continuously to help chart its interior.

The members of WET and GONG are professionals, but a similar global web draws mostly upon talented amateurs. Joseph Patterson of Columbia University established the Center for Backyard Astrophysics (CBA) a decade ago to study cataclysmic variables, binary-star systems in which an accretion disk of gas spiraling onto a white dwarf periodically unleashes energetic bursts. Patterson organizes science campaigns for the 20 or so members of his collaboration, who typically use 0.25-meter to 0.7-meter telescopes. "Although it just consists of small telescopes, the CBA is more effective than any telescope I've ever used," he said. "We focus on individual stars for hundreds of hours, which makes us very good at finding all of their rotation periods." That intensive monitoring helps the team unravel the curious wobbles and instabilities within many accretion disks, which may control when the energy spigots turn on and off.

"This is a tool that virtually nobody else has," Patterson said of his amateur assemblage. "In every country in the world, there are well-equipped amateurs interested in participating in research. They are not nuisances, and they can be tremendous assets." Janet Mattei, director of the American Association of Variable Star Observers, concurs. Her 600 observers worldwide team up to monitor the fluctuations of cataclysmic variables and other irregular objects to help guide the timing and logistics of observations by teams using Hubble, the Rossi X-Ray Timing Explorer, and other space-based instruments. "If not for amateur astronomers observing in their backyards with small telescopes, a lot of very successful observing runs with satellites would not be possible," Mattei said.

Small telescopes are the only games in town when it comes to another category of observations: bright objects and broad patches of the sky. "We do not know what is happening in the sky in real time, even at the bright end," wrote Bohdan Paczynski of Princeton recently. "This is a huge gap in astronomical research, which can be filled only with small, wide-angle instruments." That point was made clear in 2000 by Grzegorz Pojmanski of Warsaw University in Poland. Pojmanski's project, the All-Sky Automated

Survey (ASAS), unveiled nearly 3400 new variable stars in a typical slice of the Southern Hemisphere sky. His instrument was a mere 8-centimeter lens at the Las Campanas Observatory in Chile. Incredibly, just 11% of those stars were previously catalogued as variable. By that time, Pojmanski's robot had examined less than 1% of the sky, so nearly a half-million bright variable stars may await discovery. "It's sort of embarrassing to the astronomical community that the bright sky is so poorly mapped," Paczynski said.

Other established surveys also use small telescopes to tile the sky systemically. For example, the Two Micron All-Sky Survey (2MASS) uses highly automated 1.3-meter telescopes in Arizona and Chile to compile the most complete census of the cosmos to date at near-infrared wavelengths. Several ongoing programs scour space for near-Earth asteroids with 1-meter to 2-meter telescopes, such as the University of Arizona's Spacewatch and the Air Force-sponsored Lincoln Near-Earth Asteroid Research (LINEAR) program at the White Sands Missile Range in New Mexico. And when they're not zeroing in on gamma-ray bursts, ROTSE and LOTIS take images of the whole sky at least twice each night, revealing a slew of variable stars, asteroids, and other transient objects. "There's a unique new niche here for small telescopes: the time domain," said Jeff Bloch of Los Alamos, a ROTSE collaborator. "With new equipment and greater bandwidth, it will become economically feasible to do all the sky all the time."

This array of robots, networks, surveys, and innovative research makes the directors of large observatories feel that small telescopes will thrive, even though it may seem that the growing fleet of 8-meter to 10-meter glass giants will siphon away resources. Small telescopes have indeed closed at private observatories here and there. However, the most visible battles have occurred at the National Optical Astronomy Observatory (NOAO), especially at Kitt Peak in Arizona. Kitt Peak has transferred its 1-meter-class telescopes to private consortia of universities, such as the Southeastern Association for Research in Astronomy and a group led by Western Kentucky University. On the one hand, those groups have refurbished and automated the old telescopes, largely for teaching use. On the other, the instruments are no longer available to the community as a whole—especially graduate students from around the country, who did many thesis projects there.

"The small telescopes at Kitt Peak were valuable educational tools," said Kitt Peak director Richard Green. "A beginning student could get many nights." Now, many students work on data sent by remotely operated telescopes or even by studying archival data from satellites. Of all the trends involving small telescopes in the last decade or two, this is the most harmful, several astronomers told me. "We're beginning to produce students who have never been to a telescope," said John Huchra of the Harvard-Smithsonian Center for Astrophysics.

Both Green and former NOAO director Sidney Wolff acknowledged that the decisions were painful and have caused strife, but they saw no way out. "It came to a choice of broad access and educational opportunities as opposed to concentrating on excellence in forefront instrumentation and world competitive science, and we were strongly advised to take the second option," Green told me. "For years, we tried to accommodate both." That highwire act became impossible with continued erosion of the national astronomy budget during the 1990s, Wolff said. In steady dollars, NOAO's purchasing power was 40% lower in 2000 than it was 15 years earlier.

"The national observatories need to focus on large facilities that transcend the capability of single universities," Wolff stated. "The best role for small telescopes is at or near a university campus to provide student access and hands-on experience of taking and analyzing data. It's not that small telescopes produce bad science, but the kinds of forefront questions we're asking today require us to choose more powerful instruments and multi-institutional collaborations."

When one of my articles put Wolff on record with those latter words, some astronomers objected. The evidence is clear, they said, that small telescopes do indeed contribute to forefront research. From my perch on the outside, I saw broad recognition in the astronomical community that a healthy science requires both large and small telescopes. Each class has its role. Whether each receives due credit is a dicier issue. Short of a scientific poll, we won't know what the public thinks about small telescopes—although I suspect most people would regard them as the instruments of neighborhood star parties rather than important science.

But we do know where the national money is going: to the big ones. "I think small telescopes are losing ground as a result, and it pains me because they don't deserve to be," said Fred Ringwald of California State University, Fresno, the former director of the International Small Telescope Cooperative (ISTEC). "They are just as productive as ever. But to be fair to the national observatories, they simply don't have the budget to keep that many telescopes open. In order to keep providing state-of-the-art facilities, something has to give. And the small telescopes are what's giving."

This slippage sparked grumbles during my interviews. Still, Joseph Patterson of the CBA found a way to express the downer in a positive way: "Small-telescope users are passionate and militant," he said. "When you don't get the goodies that others get, it bands you together."

It also makes you imaginative about how to adapt, resourceful about how to use your time and money, and pragmatic about the role that small telescopes should play in astronomy today. I thought that Fairborn's Lou Boyd, who isn't an astronomer, put it best. "If you have a hospital lab with a CAT scanner, you also need to have people doing the blood work," he said.

"Small telescopes have their place. I don't see them going away, but I don't see them flooding the world."

ACKNOWLEDGMENT

Parts of this chapter appeared originally in Science (288: 32, 2000).

Chapter 6

Scientific Impact of Small Telescopes

Chris R. Benn, Sebastian F. Sánchez
Isaac Newton Group
Santa Cruz de La Palma, SPAIN

Abstract: We compared the scientific impacts of telescopes worldwide, on the basis of their contributions to the top-cited 1000 astronomy papers published during 1991-8. We found that during this period 1-m and 2-m telescopes together contributed about half as much highly ranked science as did all the 4-m telescopes together. This strong showing by small telescopes in the 1990's augurs well for the continued scientific impact of 4-m telescopes in the era of 8-m telescopes.

Key words: telescopes: scientific impact, scientific productivity

Some telescopes clearly have higher scientific impact than others, but there have been few attempts to quantify this, or to compare impact with cost. For example, are 4-m telescopes a better investment than 2-m telescopes? Are space telescopes as cost-effective as ground-based ones? Recently, we obtained[1] a list of the 1000 most-cited astronomy papers published worldwide during 1991-8 (the top 125 papers each year). Although high citation isn't an infallible guide to scientific impact, and citation counts are subject to a number of biases (e.g. UK/US astronomers tend not to cite foreign-language publications), most of the best astronomical research of the past decade will be represented in this sample of papers. We determined which telescopes were used to obtain the data on which each of these papers was based, and thus measured the impact (fraction of total citations generated) of each telescope averaged over 1991-4 and over 1995-8.

The impacts of ground-based optical telescopes are shown as a function of mirror diameter in Figure 1 and the 1991-4 and 1995-8 impacts are compared in Figure 2. Amongst 4-m class telescopes, CFHT leads (with the WHT in

---

[1] The citation data were obtained from the Institute of Scientific Information (ISI), Philadelphia.

T.D. Oswalt (ed.), The Future of Small Telescopes in the New Millennium, Vol. I, 49–53.

second place) but the impacts of most 4-m telescopes are remarkably similar. Keck I, in use since 1993, has an impact 8 times larger than that of typical 4-m telescopes, and this factor is similar to the ratio of collecting areas. No papers from other 8-m to 10-m telescopes (apart from Keck II, commissioned 1996) appear in the list, since most were commissioned after 1998.

The mean impact of 2-m class telescopes is a factor ~4 lower than the mean for 4-m telescopes, again consistent with the ratio of collecting areas. Comparison of individual 2-m telescopes is difficult because the numbers of papers involved are small.

The citation shares of different types of telescope are compared in Figure 3. Together, 1-m and 2-m telescopes contributed half as much science as 4-m telescopes during 1991-8. Three of the top 40 most-cited papers 1991-8 are based on data from 1-m telescopes only, including two from the microlensing survey carried out with the Mt. Stromlo 1.3-m. Extrasolar planets were also discovered using a small telescope: the Haute Provence 1.9-m (Mayor & Queloz 1995, the 9th most-cited paper of that year).

The year-2000 capital costs of 2-m, 4-m and 10-m class telescopes are

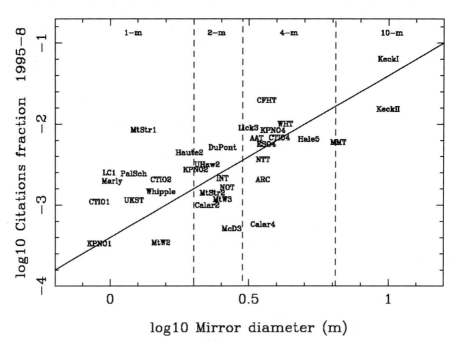

Figure 1. Citations fraction 1995-8 vs. telescope diameter for ground-based optical telescopes. Most of these telescopes were in use throughout 1991-8. The straight line indicates citation fraction = 0.6% x (diameter/4-m)^2. Statistical error N(papers)$^{-0.5}$ are ~0.2 in $\log_{10}$ for typical 4-m telescopes, ~0.3 for 2-m. In this figure, and in Figure 2 a few points have been displaced slightly to avoid overlap of labels. Digits suffixed to the abbreviations distinguish telescopes of different diameter (in meters) at the same observatory.

~$5M, 18M and 80M, respectively, while the impacts are in the approximate ratio 1:3:25 (Figure 1). If one assumes that running costs scale roughly as capital cost (Abt 1980), then the data of Figure 1 indicate that 2-m telescopes are roughly as cost-effective as 4-m telescopes. Keck I (10-m) is twice as cost-effective, but is the first of its size, and may have a bigger scientific impact than any of the 11 8-m to 10-m telescopes commissioned after 1998: 4 VLT, 2 Gemini, Subaru, LBT, HET, SALT and Grantecan.

Notably productive non-optical telescopes (Figure 2) include JCMT (about twice the impact of a typical 4-m, largely thanks to SCUBA), IRAM and the VLA. The space telescopes ASCA, BeppoSax, CGRO, COBE, Hipparcos, ROSAT all have impacts ~ 4 times higher than a 4-m telescope, but cost ~15 to 30 times as much. Comparison of the cost-effectiveness of ground-based and space telescopes is not straightforward. Some space telescopes (e.g. COBE, Hipparcos) are launched to solve a specific scientific problem which can't be tackled from the ground and they may have a short-lived community of citers, so it's not clear that citation counts are a fair measure of scientific impact. Others, such as HST, compete more directly with ground-based facilities (particularly now, with the advent of adaptive optics), and can be used to tackle similar problems, so a citation-based comparison of cost-

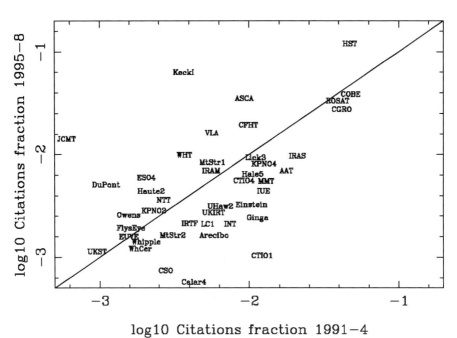

Figure 2. Citations fraction 1995-8 vs. 1991-4 for both ground-based and space telescopes. The straight line has unit slope. Only telescopes with significant impact in both 1991-4 and 1995=8 appear in the plot, e.g. BeppoSAX, Hipparcos, ISO, Keck II, RXTE and SOHO all have citation fractions > 1% in 1995-8, but zero in 1992-4.

effectiveness is fairer. HST, launched in 1990, generated 15 times as many citations as a typical 4-m telescope during 1991-8 but cost ~100 times as much, ~$2000M (much more, if the cost of servicing missions is taken into account).

For an independent measure of scientific impact, we repeated the above analysis using the 452 observational astronomy papers published in Nature 1989-98, reasoning that only papers of the highest scientific merit make it into Nature. We found a close correlation between citation fraction (above) and count of papers in Nature, except that radio telescopes are over-represented in Nature by a factor >3 relative to optical telescopes (or, perhaps, radio telescopes are under-represented in the citation counts). This discrepancy highlights the risk of incurring metric-specific biases when comparing the scientific impacts of different kinds of telescope or community.

The list of the 1000 most-cited papers can also be used to break down scientific impact by region, subject, journal or host institution (of first author). Sixty-one percent of the citations to the 1000 most-cited papers are to papers with first authors at US institutions, 11% UK, 20% European (non-UK) and 8% other (mainly Australia, Canada, Japan). Fifty-two percent of the citations are to extragalactic papers, 34% stellar/galactic, 7% solar system, 7% technical. At 4-m telescopes whose users are predominantly from North America or the UK, 75% of the cited papers (and 29 of the 34 Keck papers) are on extragalactic topics. For European (non-UK) 4-m telescopes, the fraction is 44%. The shares of the citation count by journal are: ApJ 44%, MNRAS 9%, A&A/S 10%, Nature 11%, others 29%. The most cited host institutions were GSFC, Harvard and Princeton.

For further statistics, and full details of the analysis, see Benn & Sánchez (2001). For further discussion, see Nature (2000).

The principal conclusion of our analysis is that during 1991-8, the era of 4-m telescopes, 1-m and 2-m telescopes (Figure 3) generated a substantial

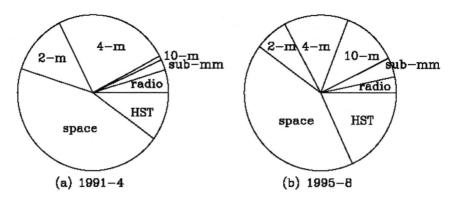

Figure 3. Citation shares of different types of telescopes, for 1991-4 and 1995-8. In this figure "2-m" includes both 1-m and 2-m class telescopes.

fraction of the science. This strong showing by small optical telescopes suggests that cutting-edge science doesn't always require the largest aperture available, and this augurs well for the continued scientific impact of small and medium-sized telescopes in the era of 8-m and larger telescopes.

REFERENCES

Abt, H.A., 1980, PASP, 92, 249
Benn C.R., Sánchez S.F., 2001, PASP, 113, 385
Mayor, M., Queloz, D., 1995, Nature, 378, 355
Nature, 2000, 408, 12
Smoot, G.F. et al, 1992, ApJ, 396, 1

Chapter 7

The Productivity of Ground-Based Optical Telescopes of Various Apertures

Helmut A. Abt
Kitt Peak National Observatory
Tucson, Arizona USA

Abstract: By scanning the papers published during the first quarter of 1996 in A&A, AJ, ApJ, *Icarus*, and MNRAS, and by noting those that were based on optical ground-based telescopes of various apertures, we learned that 82% of the papers came from telescopes with <4-m apertures; the weighted mean aperture was 2.52-m. By counting the 1996-2001 citations to those papers, we learned that 75% of the citations came from telescopes of <4-m apertures; the weighted mean aperture for the citations was 2.81-m. The mean citations per paper showed only a small dependence upon aperture. For example, telescopes of 5- to 10-m produce only twice as many citations per paper on the average as those of 1- to 2-m. Scanning the papers in the same journals during the first quarter of 2001 showed that 79% of the papers came from telescopes <4-m and the mean aperture was 2.81-m, which represent only small changes since 1996. These numbers suggest that the bulk of our ground-based optical astronomical results come from telescopes of <4-m in aperture, and that is likely to continue to be true throughout the 2001-2010 decade. The 2001 decadal astronomy survey report made no recommendations for the maintenance, new instrumentation, and new construction of those telescopes that provide most of the astronomical results derived from ground-based optical telescopes. Although commendable in other respects, that survey represents an unrealistic appraisal of contemporary optical ground-based astronomy and its needs during this decade.

Key words: telescopes: productivity, citations, publications

1.   INTRODUCTION

We wonder which optical ground-based telescope apertures are the most effective in producing the highest quality research currently being published

T.D. Oswalt (ed.), The Future of Small Telescopes in the New Millennium, Vol. I, 55–64.

and whether the plans for the remainder of this decade are realistic in supporting the construction and maintenance of the most effective telescopes. We realize that the importance of research might, at first, seem to be difficult to quantify, but a good case has been made (Abt 2000a) that large citation counts are highly correlated with papers that most astronomers recognize as being highly important. That study looked at the 53 papers selected by senior astronomers as being the most important ones published in AJ and ApJ during the 20th century. For 92% of those papers, the selected papers average 6.7 times as many citations as other papers and have 2.5 times the half-lives.

This planned study has three steps:

1. To scan all the papers published during the first quarter of 1996 in five representative major American and European astronomical journals and to count the number of papers resulting from observations obtained with ground-based optical telescopes of various apertures;
2. To count citations to those papers to determine which apertures produce the most citations;
3. To count papers published in the same journals during the first quarter of 2001 to determine whether the publication characteristics are changing rapidly during the past half decade when the telescopes of >5-m apertures have doubled in number.

2.   THE METHOD

We chose the first quarter of 1996 so that there would be five years of citations since publication. A previous study (Abt 1981a) showed that for average papers, they show a peak in citations five years after publication.

This study differs from recent studies by Abt (2000b) and Benn & Sanchez (2001) that studied only the most-cited papers. The current study includes all papers published in representative journals during a certain interval, independent of their later citation histories.

We selected the journals A&A, AJ, ApJ, *Icarus*, and MNRAS as being representative of contemporary European and American journals that publish most of the papers resulting from observations from optical ground-based telescopes. All Letters and Supplements were included. Each paper was scanned to determine whether it depended totally or in part upon ground-based optical telescopes. We included only papers quoting or using new data. Papers that stated that they used previously published data were grouped together with theoretical papers as being re-analyses. However, papers that used archival databases (e.g. IUE, ROSAT) that were not specifically used in previous studies were considered to be using new data. If a paper quoted data from two or more telescopes, partial credit is given to each. If a journal of observations is included to show the fraction of data coming from various telescopes, then those fractions were used to assign partial credits. For papers depending partly upon new ground-based data and partly upon ground-based radio or space data, credit was again prorated. Hence the totals show many fractional papers and there are some round-off errors.

The telescope apertures were binned as 0- to <1-m, 1.0- to <2-m, 2.0- to <3-m, etc. These are referred to briefly as 0.5-m, 1.5-m, 2.5-m, etc. in the tables and figures. The telescope apertures of ≥5-m in the 1996 publications included only the 5.1-m Hale, the 6-m Special Astrophysical Observatory, and the 10-m Keck I Telescopes. The weighted (by numbers of papers or partial papers) mean aperture for those was 7.0-m, so the data were plotted with that abscissa.

The citation counts to those 480 papers or partial-papers were found in the on-line Web-of-Science of the Science Citation Index at the University of Arizona's Science Library on 2001 June 23. The counts were inclusive for the five years prior to that date. Again, papers partially attributable to individual ground-based optical telescopes were given partial citation credits.

3.   RESULTS AND CONCLUSIONS

The numbers of papers from ground-based optical telescopes in the five representative journals are given in Table 1. Where there were sufficient data, Letters papers did not give significantly different results than main-journal papers. The last column of Table 1 gives the weighted mean aperture for each journal. The last row of Table 1 shows the percentages of papers attributable to various aperture telescopes. Figure 1 is a plot of the summed results for all five journals.

Of course the individual ordinates in Figure 1 (and in later figures) depend upon how *many* telescopes there are of a given size, e.g. if there had been no 1- to 2-m telescopes, that ordinate would be zero. However what Figure 1 shows is that with the contemporary complement of telescopes, the bulk of the published papers come from telescopes of <4-m in aperture.

We obtain the following results:

1. The mean apertures do not differ substantially among the five journals, ranging only from 2.24- to 2.94-m. The mean is 2.52-m.
2. Telescopes of <4.0-m aperture contribute 82% of the papers.

However, not all papers are equivalent in importance, so what is more significant are the numbers of citations produced in five years by those 332 papers. Those numbers are given in Table 2, where the weighted mean apertures are given in the last column. The sums are plotted in Figure 2.

Table 1.
Numbers of 1996 Papers from Various Ground-based Optical Telescopes

Journal	Telescope Apertures (meters)						Sum	\<A\>
	0.5	1.5	2.5	3.5	4.5	7.0		
A&A	17.9	46.9	13.2	24.1	7.8	4.5	114.3	2.24
AJ	14.5	24.6	11.6	8.1	8.7	5.9	73.4	2.37
ApJ	10.2	18.9	16.4	20.4	14.8	9.6	90.3	2.94
Icarus	1.0	1.0	0.3	0.0	0.3	0.3	3.0	......
MNRAS	4.0	16.1	8.8	13.1	9.0	0.0	51.0	2.64
Sum	47.6	107.5	50.3	65.7	40.6	20.3	332.0	2.52
%	14.3	32.4	15.2	19.8	12.2	6.1	100.0	

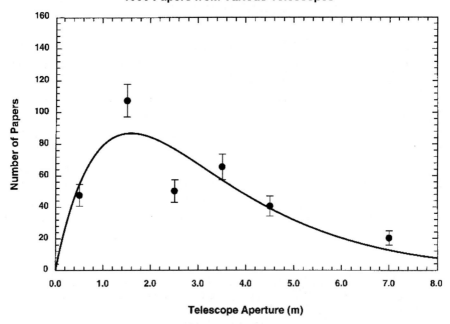

Figure 1. The numbers of papers based on ground-based optical telescopes in A&A, AJ, ApJ, Icarus, and MNRAS during the first quarter of 1996 are shown as a function of telescope aperture. The bins are 0- to < 1-m, 1.0- to <2-m, etc., except that the three contributing telescopes >5-m have a weighted mean aperture of 7.0-m. The standard errors are the square roots of the counted numbers.

Table 2.
Citations During 1996-2001 to 332 Papers Resulting from Ground-based
Optical Telescopes of Various Apertures

Journal	Telescope Apertures (meters)						Sum	<A>
	0.5	1.5	2.5	3.5	4.5	7.0		
A&A	130.4	417.8	149.7	330.4	61.0	47.0	1136.2	2.43
AJ	181.1	340.5	161.9	92.7	117.8	143.0	1037.0	2.55
ApJ	128.5	465.0	292.9	402.5	320.6	329.9	1939.4	3.18
Icarus	4.0	4.0	0.7	0.0	0.7	0.7	10.1	.....
MNRAS	35.5	154.5	134.4	144.0	145.5	0.0	613.9	2.84
Sum	479.9	1381.8	739.6	969.6	645.6	520.6	4737.1	2.81
%	10.1	29.2	15.6	20.5	13.6	11.0	100.0	

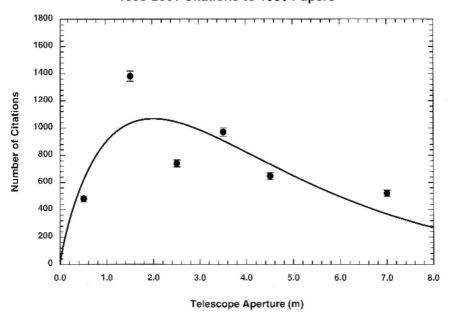

1996-2001 Citations to 1996 Papers

Figure 2. The total citations during 1996-2001 June 23 to the papers counted in Figure 1
are plotted as a function of telescope aperture. The binning and standard errors are
computed as in Figure 1.

The results from these are:

1. Again the range in mean apertures is not large, namely 2.43- to 3.18-m.
2. The average for the five journals is a mean aperture of 2.81-m, not substantially larger than the mean aperture of 2.52-m obtained by counting papers.
3. Telescopes of <4-m in aperture contribute 75% of the citations.

Figures 1 and 2 are similar in shape (which is fit by a Hoerl model), implying that the numbers of citations per paper are not a steep function of telescope aperture, A.   The ratio of the two plots is shown in Figure 3, where the best fit is given by: citations per paper = 8.61 + 2.16A.   For example, a typical paper from a 7-m telescope produces only twice as many citations as a typical paper from a 1.5-m telescope.   This result may be very surprising to many people who are under the impression that a large telescope is dramatically more effective in producing important astronomical results than a small telescope.

There are several reasons for the small dependence of mean citation counts on telescope aperture.   (1) Perhaps the ingenuity of the astronomer is as important as the sensitivity of his/her equipment.   (2) With smaller telescopes astronomers are more likely to explore and experiment.   Then after they have

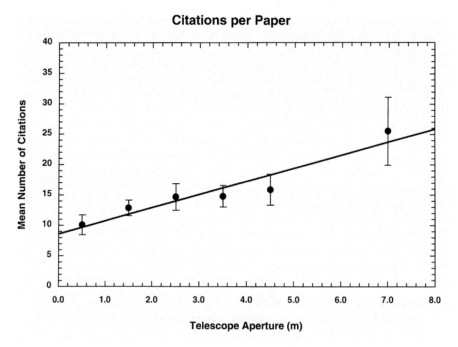

Figure 3. The average citations per paper are computed from the data in Figures 1 and 2. The standard errors are the root-mean-square combined errors from Figures 1 and 2.

discovered an important new result, they can qualify for large telescope time for follow-up studies, but the small telescopes have gotten the credit for the initial discoveries. We think of the work on rotation curves in galaxies, which was done with 0.9-m telescopes that showed the existence of missing mass. The discovery of the Lyman-α forests was done with a 2.1-m telescope. The first extraterrestrial planets were found with 1- and 2-m telescopes. The first gravitational lens was found with a 2.1-m telescope. In each of these cases larger telescopes were in use at the time. (3) Small telescopes are often used for fundamental calibrations, such as Landolt's photometric standards measured with 0.9- and 1.5-m telescopes. (4) Schmidt and other small telescopes are used for surveys that discover interesting objects, such as the Sloan Digital Survey being done with a 2.5-m telescope.

Considering that the construction and maintenance costs of optical ground-based telescopes vary approximately as 2.4-power (Abt 1980), a 7-m telescopes costs 40 times as much as a 1.5-m telescope, but only produces twice as many citations per paper. Of course there are projects that can be done only with large telescopes, but, for example, most observing done on stars brighter than 10th magnitude can be done much more economically with a smaller telescope.

These results refer to the papers published in 1996 and are based on telescopes that were available several years before the publication dates. The three large telescopes (mentioned above) of $\geq$5-m in aperture that produced the 1996 papers have since been augmented by the 8.2-m Very Large Telescope, 9.2-m Hobby-Eberly Telescope, and 10-m Keck II Telescope. Have the results quoted above for the 1996 papers changed drastically with the new use of twice as many telescopes of $\geq$5-m aperture?

We scanned papers in the same five journals for the first quarter of 2001. The data from the ground-based optical telescopes are given in Table 3. The entries are directly comparable with those in Table 1. For instance, the 35.7 papers coming from the six 5- to 10-m telescopes is nearly double the 20.3 papers from the three 5- to 10-m telescopes in 1996.

We derive the following results:

1. The weighted mean aperture was 2.81-m, not substantially greater than the 2.52-m in 1996, despite a doubling of the number of telescopes >5-m.
2. Telescopes of <4-m aperture contributed 79% of the papers, not substantially less than the 82% in 1996.
3. Because of the similarity of paper counts and citation counts shown in Figures 1 and 2, we can expect that the eventual citation counts for the 2001 papers will again show that the bulk of the citations will come from telescopes <4-m in aperture.

Table 3.
Numbers of 2001 Papers from Various Ground-based Optical Telescopes

Journal	Telescope Aperture (m)						Sum	<A>
	0.5	1.5	2.5	3.5	4.5	7.0		
A&A	18.2	36.9	30.4	14.7	5.5	10.0	115.7	2.48
AJ	15.2	17.2	13.2	10.8	8.9	8.8	74.1	2.78
ApJ	13.7	16.8	17.1	23.7	14.5	16.6	102.4	3.31
Icarus	0.0	4.2	1.2	2.5	0.0	0.0	7.9	.....
MNRAS	4.2	6.9	4.5	7.7	3.3	0.3	26.9	2.51
Sum	51.3	82.0	66.4	59.4	32.2	35.7	327.0	2.81
%	15.7	25.1	20.3	18.2	9.8	10.9	100.0	

4. The slow changes in five years suggest that throughout the 2001-2010
    decade the bulk of the papers and citations from optical ground-based
    telescopes will still come from telescopes <4-m.

It may be of interest to know what fractions of the current papers come
from ground-based optical telescopes. Therefore in scanning the 1503 papers
published in the five journals during the first quarter of 2001, we noted for
each one the instrumental sources of the data. Many papers were credited to
two or more instruments, in the same manner described earlier. A category
labelled "theory" included both analytical and numerical theories, modelling,
re-analyses of data discussed earlier, and (a very minor addition) theoretical
and experimental atomic and molecular measurements. There is some
confusion between space ultraviolet and space optical, as was done by the
Hubble Space Telescope, because the dividing wavelength is unclear; the two
might well be combined. Also, in an interval as short as three months, the
publication of a large set of papers from a new instrument (e.g. 56 papers
from XMM-Newton in the January A&A Letters) will give non-typical high
values for one field. The subdivisions of papers or partial papers are given in
Table 4. We see that about 45% of the papers are theoretical, which
continues the century-long trend that started from 5% in 1910 (Abt 1981b).
About 22% of the papers are based on optical ground-based telescopes, and
about 10% each from ground-based radio and space X-rays; the remaining
10% come from all other equipment.

Table 4. Subdivision by Instrumentation

Equipment	Number of Papers by Journal					Sum	%
Ground-based:	A&A	AJ	ApJ	Icarus	MNRAS		
theory	173.5	31.0	345.0	30.0	101.0	680.5	45.3
γ, particles	2.0	0.0	4.0	0.0	0.0	6.0	0.4
optical	115.7	74.1	102.4	7.9	26.9	327.0	21.8
radio	44.4	19.0	69.6	3.0	11.3	147.3	9.8
Space:							
γ	1.0	0.0	7.0	0.0	1.0	9.0	0.6
X-rays	76.2	2.3	90.2	0.0	11.0	179.7	12.0
UV	13.9	5.8	40.0	1.5	1.3	62.5	4.2
optical	6.0	8.5	18.4	0.5	4.3	37.7	2.5
IR	22.8	5.5	14.5	4.0	1.0	47.8	3.2
radio	2.0	0.0	2.5	0.0	1.0	5.5	0.4
Sum	457.5	146.2	693.6	46.9	158.8	1503.0	100.2

The National Academy of Sciences commissioned a new decadal survey of astronomy and astrophysics (McKee & Taylor 2001). Among its recommendations for ground-based optical astronomy are the construction of large (>6-m) and very large (>10-m) telescopes, and a telescope system instrument program for "large" telescopes. The report is silent about the maintenance, improvement, and new construction of telescopes of <4-m except to leave those up to individual universities and consortia. In other words, if your university does not have a 2- or 3-m telescope and is not a member of a small consortium that owns one, you are out of luck if the funding for the national centers (KPNO and CTIO) continues to decline so that the remainder of their small telescopes have to be closed. While we do not deny the importance of large telescopes for studying faint objects, the survey's lack of recommendations for the upkeep and improvement of those telescopes that are currently providing the bulk of the ground-based optical astronomical results is difficult to understand. Its survey of contemporary optical ground-based astronomy was inadequate and unrealistic.

4.   REFERENCES

Abt, H. A. 1980, PASP, 92, 249
Abt, H. A. 1981a, PASP, 93, 207
Abt, H. A. 1981b, PASP, 93, 269
Abt, H. A. 2000a, Scientometrics, 48, 65
Abt, H. A. 2000b, BAAS, 32. 937
Benn, C. R., & Sanchez, S. F. 2001, PASP, 113, 385
McKee, C. F., & Taylor, J. H., Jr, 2001, ed. *Astronomy and Astrophysics in the New Millennium* (Washington: National Academy Press)

Chapter 8

The Future Operation of European Medium-Sized Telescopes: The Situation at Calar Alto Observatory

Roland Gredel
Max Planck Institut für Astronomie
Heidelberg, GERMANY

Abstract: Various recent user inquiries show that a significant need exists to maintain medium-sized telescopes (2- to 4-m class) in the future, operating hand in hand with the 8- to 10-m telescopes. First, a large number of scientifically important programs are performed more adequately and more efficiently at the medium-sized telescopes. Examples are large field imaging and spectroscopy, studies on brighter objects, long term monitoring and survey programs and support of surveys at other wavelengths (X ray, radio), and accurate astrometry and photometry of reference objects and unknowns. Second, the growing complexity of instrumentation requires telescope time for instrument development and optimization that is more economically available at the medium sized telescopes. Third, astronomy as an international scientific venture that stimulates and inspires future technological enterprises requires a significantly increased share of telescope time for training and education. And fourth, nationally operated medium-sized telescopes will play an important role in integrating European astronomical research, by opening up the national programs to researchers throughout Europe.

Key words: Calar Alto observatory, medium-sized telescopes, OPTICON

1. WHO MOVED MY CHEESE?

If You Do Not Change, You Can Become Extinct (Haw to Hem, in S. Johnson, 'Who Moved My Cheese'). What is a general truth about an intelligent reaction to changing conditions in life probably applies to astronomical observatories as well. It is arguably a correct statement that the most advanced astronomical research projects have shifted from the 4-m to the 8-m class telescopes. This is a profound change and it has prompted a

T.D. Oswalt (ed.), The Future of Small Telescopes in the New Millennium, Vol. I, 65–83.
© 2003 *Kluwer Academic Publishers. Printed in the Netherlands.*

number of user inquiries and assessments concerning the future role of medium-sized (2- to 4-m) telescopes. In Europe, it was not clear *a priori* whether there was a scientific need to continue the operation of medium-sized telescopes at all, at least to those who concentrated their research to the southern hemisphere. But also, operational implications that may result from significantly decreased funding have to be discussed, together with the general policy of the European Community to improve structural weaknesses for researchers in Europe.

Scientific interest: The national astronomical communities in Europe have expressed a clear scientific need to continue the operation of 2- to 4-m class telescopes for at least one more decade. The 4-m class will remain an important research tool in the future. The situation may be similar to that which existed during the last three decades, when the 1- to 2-m class telescopes provided forefront science at a time where a total of eight 4-m class European telescopes became available. This is the same as saying that size is not all that matters (e.g. Hellemans 2000). Even now, fundamental astronomical discoveries are being made by use of telescopes of 1- to 2-m aperture. Examples are MACHOs (massive astrophysical compact halo objects) and extra solar planets.

Operational costs: The largest national observatories in Europe are the Isaac Newton Group of Telescopes (ING) on La Palma funded by PPARC and the Calar Alto Observatory in Spain funded by the Max Planck Gesellschaft/Max Planck Institut für Astronomie (MPIA). Both observatories are unsurpassed for the range and flexibility of their facilities. At both sites, the ever-increasing needs of the national astronomical communities have been unconditionally fulfilled, which has led to the costly maintenance and operation of a wide range of general-purpose instruments that can sing and dance. The anticipated entry of the UK to ESO and the participation of the MPIA at the Large Binocular Telescope (LBT) will require a more cost effective operation of the two observatories in the future. The need to free resources will apply for other national observatories in Europe as well, such as the Nordic Optical Telescope (NOT) on La Palma, the Observatoire de Haute Provence (OHP), the Observatoire de Midi Pyrenees on Pic du Midi, and possibly, the Telescopio Nazionale Galileo (TNG) on La Palma, if only after a few years for the latter.

General European Policy: In general, the European Community (EC) is suffering from structural weaknesses when research is concerned, compared to the United States and Japan. Generally, there is insufficient money and manpower available to research. In 1999, the EC member states spent an average of 1.9% of their gross domestic product on research, compared to 2.6% for the USA and 2.9% for Japan. From the total workforce, 0.53% are in research in the EC, compared to 0.74% in the USA and 0.89% in Japan. There is the recognition that research is a central component of the knowledge-based economy and society that is developing worldwide. As far as European astronomy is concerned, access to state of the art 2- to 4-m class telescopes and their instrumentation is largely restricted to researchers from a few privileged nations only. In order to level out these disadvantages (depending on the point of view), the EC encourages Member States and EC Associated States to step up the coordination of research activities which are carried out in Europe. In particular, the EC is funding projects which open up national programs, and which provide access to national facilities to researchers throughout Europe. In promoting the European Research Area (ERA), the EC supports the participation of EC candidate countries, with the objective to open ERA to Mediterranean third (non member) countries, to Russia and the states of the Commonwealth of Independent States (Armenia, Azerbaijan, Belarus, Georgia, Kazakhstan, Kyrgyzstan, Moldova, Tajikistan, Turkmenistan, Ukraine, Uzbekistan). It will be discussed in Section 5 below how a network of nationally operated telescopes may benefit all European researchers.

The following is an attempt to summarize discussions among national telescope operators in Europe which aim to find ways to adapt to the changes which are imposed by arrival of the large telescopes, the need to more efficiently operate the national observatories, and the political aim to make their telescopes more universally available to researchers throughout Europe. In particular, the change from the costly, multi-purpose national observatory to one which operates specialized telescopes embedded in a European coordination network under the umbrella of OPTICON, the coordination network in European optical and infrared astronomy, is outlined in Section 5. As an example, the changes that are under way at the Calar Alto Observatory are described in detail in Section 2. Changes at other national observatories operated by European agencies are summarized in Section 3. The co-added values and synergism which emerge from a coordination of European national observatories are pointed out, together with a future role of medium sized telescopes in training and raising the awareness in scientific and technology matters among the young generation.

2. THE CALAR ALTO OBSERVATORY

The Calar Alto Observatory was founded in the 1970s in order to provide astrophysicists at the Max Planck Institutes and Universities in Germany with a modern observing facility in the northern hemisphere. Since then, the Calar Alto Observatory together with the LaSilla Observatory in Chile, has played a crucial role in putting German ground–based optical and infrared astronomy in an internationally competitive position again.

As the annual reports of the Max Planck Institut für Astronomie (MPIA) show, many exciting results have been obtained at Calar Alto's 3.5-m, 2.2-m, 1.23-m, and Schmidt telescopes and published in international journals. Moreover, expertise in the construction of modern instruments and observational techniques was developed, at both Max Planck Institutes and Universities which allow German consortia to contribute in a major way to the recent development of instruments for 8-m class telescopes, most notably the Very Large Telescope (VLT) of the European Southern Observatory (ESO). Telescope time is awarded semi annually via peer reviewed proposals. The typical time allocation is about 40% for MPIA, 45% for the remaining German community, and 10% for Spain. Weather patterns are similar to southern California and Arizona, and the number of photometric and clear nights does not differ from those at Kitt Peak, Mt. Graham, Mt. Palomar, and Lick observatories.

2.1   Calar Alto Site Characteristics

Photographic star trails and photoelectric knife edge seeing measurements obtained in the early 1970s during a total of 171 nights of site testing on Calar Alto suggested that the median value of the seeing is 0.57". (Birkle 1976). Calar Alto has been operating a differential image motion monitor (DIMM) since March 2001. Statistically significant results are not available yet but existing measurements indicate that the median seeing is better than 0.8" indeed. During the summer months, Calar Alto suffers from occasional and significant enhancements in the extinction due to dust from the Sahara desert. During the last 15 years, an average of 10 nights per year on Calar Alto are affected by excessive extinction. This compares favorably with the situation on the Canary Islands, where some 25% of the July–September period suffers from high extinction from Saharan dust (Guerrero et al. 1998). The wavelength dependent extinction on Calar Alto varies significantly and is not described in terms of a standard extinction curve. An extinction monitor is presently being set up at the observatory, with the goal to provide extinction values every night to the observers. Measurements of the extinction in the near infrared wavelength region show that extremely transparent nights occur on Calar Alto, mostly during the winter months. Extinction values as low as 0.1 magnitude per airmass have been derived in the K' band. Such low values compare with those of Mauna Kea.

Weather losses altogether amount to some 35-40% of the time available to the observations. Most critically, a large number of programs are affected by clouds throughout the year. What makes things worse is a severely degraded seeing at the 3.5-m telescope, where the median value is 1.1". This is in conflict with the superb optical quality the telescope itself offers (it is Zeiss, which is nice, but it was twice the price). The frequent and intermittent weather downtime, and the bad seeing at the 3.5-m telescope, together with the absence of technical support during weekends in the past, resulted in a relatively low scientific production rate compared to international standards, both in terms of papers produced per night of observation and citations per paper. A program was initiated in early 2000 to significantly increase the efficiency of observations on Calar Alto. It is described in detail in Section 2.4.

2.2 Calar Alto Instrumentation

Modern instrumentation exists on all telescopes. The 3.5-m adaptive optics system ALFA is state of the art. The 3.5-m near infrared prime focus camera OmegaPrime, with a field of view (FOV) of 7', has been in operation for the last five years. It is being replaced by the NIR wide field imager Omega2000 in spring 2002, which will provide a FOV of about 15'. LAICA, an optical wide field imager with 1 square degree of FOV, is presently being commissioned at the 3.5-m prime focus. PMAS, the Potsdam Multi Aperture Spectrograph, provides integral field spectroscopy using a 16x16 lenslet array, with a spectral coverage from 350–900 nm in one shot at spectral resolutions of 1.35–0.45 Å/mm. Workhorse instruments at the 3.5-m telescope include the focal reducer MOSCA, the double spectrograph TWIN, and OmegaCass, a 1K Hawaii NIR instrument with imaging, spectroscopic, and polarimetric capabilities. BUSCA, a 4-color simultaneous CCD camera that permits sub 0.01 mag. color photometry has recently been commissioned at the 2.2-m telescope. Other instruments offered at the 2.2-m include the optical fiber fed echelle spectrograph FOCES, the NIR camera and spectrograph MAGIC, and the focal reducer CAFOS. A high resolution Coudé spectrograph designed for the optical wavelength region but which may also be used in the NIR is available as well. Among the various guest instruments that are used regularly are fast photometers at the 3.5- and 2.2-m, a 3D NIR camera at the 3.5-m, and SCIDAR at the 1.23-m telescopes. Detailed information about the Calar Alto telescopes and instrumentation is available at http://www.caha.es and http://www.mpia-hd.mpg.de. About a dozen instruments are routinely offered at the 3.5-m and 2.2-m telescopes altogether. Each of these instruments requires support from Calar Alto and partly from MPIA, in particular to maintain and upgrade the software, the electronics and hardware, and the optics. It must not be stressed that such a huge instrument park is expensive both in terms of operation and maintenance. A way to decrease the instrumentation available on Calar Alto and at other national observatories, which will not compromise the astronomical research, are described in Sections 5 and 5.1.

2.3   The Calar Alto User Inquiry

In order to address the future needs of the German and Spanish communities on Calar Alto in the era of the 8-m telescopes, a user survey was carried out in 1999 on the world wide web page of the MPIA. The following questions were addressed:
– How will the interest in – and the need for – Calar Alto change within the German and Spanish community with ever increasing access to 8- to 10-m telescopes?
– Is it scientifically and financially justified to continue operation of Calar Alto in the era of the new 8- to 10-m optical and infrared telescopes?
– In which scientific fields can the Calar Alto telescopes remain competitive and scientifically productive during the next decade(s)?
– What are the technical prerequisites for a scientifically successful and productive continuation of Calar Alto?
– What are the optimal modes of operation of the observatory in the future (guest observers, service observing, dedication to very large projects, remote observing)?

From the replies received, there was unanimous consensus that Calar Alto will still be essential for the competitiveness of German, but also Spanish, astronomy in the near to medium term future, independent of whether the user will or will not have access to an 8-m class telescope. Users with access to 8–m class telescopes considered a ratio between observing times at 8-m vs. smaller telescopes of 1:1 to 1:5 most likely. Smaller telescopes are needed for monitoring projects, surveys, and preparation of observations at 8-m class telescopes. Only a minor fraction of users thought that the importance of Calar Alto may decline after the first decade of the 21st century.

The high risk of bad weather, minimal technical support during weekends, and the degraded image quality of the 3.5-m telescope were identified as the key problems which prohibit a more efficient use of the Calar Alto Observatory. The strengths of the Calar Alto Observatory were seen in the (partly unique) infrared instruments and adaptive optics. But also workhorse type instruments such as the TWIN spectrograph or CAFOS were valued highly. There was no general consensus on which instruments could be taken out of operation. Continuing cutting edge instrument development was seen as essential for the future success of Calar Alto. Trading of observing time with other observatories was considered a possibility to gain access to instruments not available at Calar Alto. The users welcomed large projects and were willing to join such projects if these are of value for their own research, but they were strictly against forced marriages. Large projects should be selected on the basis of scientific merit, uniqueness, optimal use of instruments and broad relevance. Data from large surveys should be archived if feasible.

The recommendations of the Calar Alto Advisory Committee to the MPIA directorate summarize the users' views that Calar Alto does fulfill a crucial role in the era of the 8-m class telescopes. This role is different from its previous one but vital for the success of German and Spanish astronomy in the future. Calar Alto has developed a leading position in NIR imaging and low resolution spectroscopy and should stay competitive in these fields. The same is true for adaptive optics. Calar Alto telescopes will remain essential for:
– survey work to find and define targets for observations with 8- to 10-m class telescopes
– monitoring of variable sources
– systematic investigation of known objects with large samples, and the discovery of rare objects or new classes of objects
– optical/NIR identification and follow up work of space based and radio observations
– building and testing of novel instruments as fast reactions on new technological developments

2.4   The Future Operation of the Calar Alto Observatory

In the beginning of 2000, an aggressive plan has been implemented to improve the scientific productivity of the Calar Alto observatory. Full coverage of technical support during weekends has been implemented in January 2000. The two major shortcomings of Calar Alto, significant losses due to bad weather and degraded seeing at the 3.5-m telescope, are being addressed by the introduction of service observations and flexible scheduling, and a modification of the 3.5-m dome. A very novel aspect of the future operation of Calar Alto which may arise from an integration into a European coordination network is described in Section 5 below.

In order to compensate for weather losses, service observations are being offered since the second semester of 2000. The quasi-guaranteed completion of accepted programs in service mode is an important step towards a significantly improved scientific output rate. Flexible scheduling or service mode A has been implemented for highly rated scientific programs which requested ALFA together with the laser guide star in the summer of 2000. During the last two semesters, additional, highly rated programs were carried out in service mode A as well. Other programs were scheduled in service mode B, where observations are obtained by Calar Alto staff at fixed times. As for regular visitor observations, there is no compensation of weather losses for category B. A total of some 15% of all programs at the 3.5-m and 2.2-m telescopes are presently being carried out in service mode. During the next 1-2 years, service observations will be increased to a total fraction of some 60-

70%. This should be possible without a significant increase of the staff once a joint control room for observations at the 3.5-m and 2.2-m is available.

In addition to service observations, weather losses are being compensated by allocating very large fractions of telescope time to long term survey projects. Four surveys are presently being active on Calar Alto, CADIS, MUNICS, follow up studies of the Sloan Digital Sky Survey, and WeCAPP. The surveys are briefly described: CADIS, the Calar Alto Deep Imaging Survey, is an extragalactic key project at the MPIA conducted on Calar Alto (http://www.mpia-hd.mpg.de/galaxies/cadis/science_index.html). The survey combines a very deep emission line survey with a limiting line flux of F_{lim} = 3×10^{20} W m^{-2} with a deep multicolor survey in 15 broad and intermediate bands between 400-2100 nm, with limiting magnitudes of B= 25, I= 23, K= 20. The observations are carried out in seven extragalactic fields of about 120 arcmin^2 each. Scientific goals are the study of the high luminosity tail of Lyman alpha bright primeval galaxies, the luminosity function of field galaxies and its evolution since z = 1, the global star formation rate since z = 1.2, the evolution of large structures since z = 1.2, a census of quasars at high redshift, a census of extremely red objects, and the stellar content and the structure of the Galaxy.

The Sloan Digital Sky Survey (SDSS) (http://www.sdss.org) is conducted at a specialized 2.5-m telescope of the Apache Point Observatory supported by a 0.5-m photometric telescope which monitors the atmospheric conditions during the survey. Researchers at the MPIA have access to the SDSS, and follow up observations are being carried out on Calar Alto since the first semester of 2001. The large variety of follow up observations with many different scientific aims often require a few hours of observations only. The data are conveniently being collected through service observations.

The Munich Near IR Cluster Survey (MUNICS), lead by the extragalactic group the University Observatory of Munich, is a wide field photometric survey of QSOs with redshifts of 0.5 < z < 2 selected in the K' band. Optical follow up observations are carried out to obtain photometric redshifts. The survey covers a field of roughly one square degree. It is medium deep, QSO candidates have K'$_{lim}$ = 19.5 magnitude. Scientific aims are the identification of clusters of galaxies at high redshift by detecting their luminous early type galaxy population, the study massive galaxies at high redshift, and the evolution of galaxies in high density regions as a function of redshift. Weather losses during regularly scheduled observations have not allowed for an optimal use of the data. A minimal investment of a few nights of service observations has provided complementary optical or NIR data which were missing previously. This has increased the data which is now useful to the survey by a factor of 2.

WeCAPP, the Wendelstein Calar Alto Pixellensing Project, is a micro lensing study in M31 and aims to detect MACHOs, or baryonic dark matter in the galactic halo. WeCAPP is similar to microlensing studies in the Small and Large Magellanic Clouds and in the Galactic Bulge and complements other pixellensing studies such as the POINT AGAPE survey in M31 (http://www.point-agape.org), or the AGAPEROS application of pixellensing of a small subset of EROS 1 microlensing observations (e.g. Melchior et al. 1998). WeCAPP is carried out on the Wendelstein 0.8-m telescope and the Calar Alto 1.23-m telescope. During the last three semesters, the project has received some 30% of the total time available at the 1.23-m telescope. The observations have initially been gathered in service mode. They are now being carried locally out by members of the research group which stays on Calar Alto for several weeks.

The immediate compensation of weather losses is an important aspect to increase the efficiency of Calar Alto. Equally important is the elimination of local effects that degrade the seeing at the telescopes. At the 3.5-m telescope, dissipation of energy in the 6-floor, 43-m high building, together with a very poor ventilation during observations, result in a degradation of the seeing by some 0.3-0.4". A major step to eliminate the local effects is under way since July 2001. During the three summer months of 2001, the telescope is closed down for technical work at the dome. The former constructor, Dillinger Stahlbau of Saarlouis, Germany, has signed a contract worth 1.2 million German marks to install a dome ventilation system. It consists of automatic flaps in the dome structure that will provide a net opening of some 20 m^2. The work is in progress at the writing of this article. Once operational in October 2001, the dome vents will provide very efficient natural ventilation during observations, and remove all local temperature differences that are considered as the main cause of the seeing degradation. The dome modifications are similar to those at UKIRT, the Kitt Peak 4-m Mayall telescope, and the CTIO 4-m Victor Blanco telescope.

A number of additional measures have been taken to control the thermal environment of the telescope and the building. Electronic workshops which were located in the 3.5-m building have been relocated, followed by decreased heating in all floors. In order to minimize seeing degradations that arise from mirror seeing, the mirror cell is being actively cooled during daytime by means of a flow of cool air. The 3.5-m primary mirror is rather thick, and does not react quickly enough to changes in the ambient air temperature. The cooling during daytime also avoids a temperature increase from heat that is produced by the electronics of instruments mounted in the Cassegrain focus and other local heat sources. The electronics of the new instruments such as Omega2000, LAICA, and PMAS, will be placed in a closed box, with all heat extracted from the telescope. Future measures to stabilize the thermal environment of the dome and the telescope will include a very restricted access to the building and the removal of the control room which is presently located near the telescope at the 4th floor. A local control room will be installed in the first floor in the near future. It is expected that the largest fraction of observations in the future will be carried out from a joint control room for the 3.5-m and 2.2-m, which is presently being built some 500-m away from the telescopes.

Improved efficiencies will also result from an optical engineering program which has been implemented on Calar Alto. The reflectivity and the scattering of the mirrors of all telescopes is constantly measured and monitored. Goals are to maintain the reflectivity of the mirrors above 87% at any given time. An inflatable rubber ring has recently being delivered by Deutsche Schlauchboot GmbH. It allows for a waterproof sealing of the mirror cell and washing of the mirror *in situ*, without removing it from the cell. Washing of the mirror will be performed in intervals of 2-3 months. It will be complemented by high and low pressure CO_2 cleaning (depending on humidity) once a week.

2.5   Automated and Robotic Telescopes

Plans are presently being developed to automate the 1.23-m telescope, possibly converting it to a semi-robotic telescope altogether. In the future, a very large fraction of time may be allocated to the 1.23-m telescope for follow up studies of the Hamburg Quasar Survey (HQS). The HQS was taken in the 1985–1997 Period using the Calar Alto Schmidt telescope equipped with an objective prism. The HQS archive contains some 1300 plates on Kodak IIIa J and provides a flux limited sample of 415 quasars with B < 17 in an effective area of 3700 square degrees. The astrobiology department of INTA/CSIC (Instituto Nacional de Tecnica Aerospacial) has proposed the installation of a 0.5-m fully robotic telescope on Calar Alto. Scientific drivers relevant to astrobiology are the photometric detection of extra solar planets, follow up studies of solar type stars, and astrometry of near Earth objects.

Future observations on 1- to 2-m class telescopes may rely on fully robotic telescopes altogether. An example is provided by the Liverpool telescope (http://telescope.livjm.ac.uk) which is presently under construction on La Palma in Spain. The normal operation of the Liverpool telescope will not require local or remote supervision. A major advantage of the robotic telescope is its fast response time to e.g. target of opportunity observations. The telescope control system will handle all tasks related to scheduling, start up and closing, weather monitoring, sequencing within observations, and error handling and reporting. In short, it will handle all tasks a human observer struggles with during a regular observing run. There are some sixty 2-cm to 2-m automated telescopes in operation or under construction around the globe. Among the larger telescopes are the 2.1-m Tennessee Spectroscopic

Survey Telescope, the 1.3-m Mt. Stromlo telescope dedicated to the search of Trans Neptunian Objects, STELLA, the 1.2-m telescope of Hamburg, AIP, and IAC on Tenerife, the 1.25-m SpectraBot of Indiana University, IRAIT, the 0.8-m Italian robotic Antarctic infrared telescope, BAIT, the Leuschner Berkeley Automated Imaging Telescopes with 0.5-m and 0.76-m aperture, the 0.8-m Katzman Automatic Imaging Telescope at Lick, the 2 x 0.76-m Wolfgang Amadeus telescopes operated by the University of Vienna in Patagonia and Arizona, and the 0.8-m Catania Automatic Photoelectric Telescope in Mt. Aetna. Possibly the smallest among all was the 0.02-m telescope of ASAS, the All Sky Astronomical Survey at Las Campanas which was in operation from 1997–2000. More sophisticated is ROBONET, a global network of six 2-m robotic telescopes, recently proposed to the UK government but rejected for the time being.

3. ASSESSMENTS BY THE BRITISH, ANGLO AUSTRALIAN, CANADIAN FRENCH, AND EUROPEAN COMMUNITIES

The conclusions of user surveys on the future role of medium sized telescopes by the British, Anglo-Australian, Canadian French, and European communities, are summarized. Note that the situation at other national observatories which operate medium sized telescopes may be different from that for Calar Alto, as the focus of the astronomical research may differ from community to community.

3.1 PPARC and the Isaac Newton Group of Telescopes

The UK has recently completed a comprehensive long range planning process in astronomy and particle physics. The Ground Based Facilities Committee (GBFC) calls for the UK community to have access to, in each hemisphere, up to 50% of each of an 8-m telescope, a wide field 4-m telescope and a general purpose 4-m telescope, plus a 10% share of the Large Millimeter Array ALMA. The international review panel believes strongly that there will be a major role for medium sized telescopes in the era of 8-m telescopes. However, the ING on La Palma will have to be operated more cheaply in the future and be equipped with excellent instrumentation which is reliable, and which exploits the strengths of the telescopes and site. It was suggested that the ING develop a viable long term instrumentation plan, soundly based on astronomical objectives. Two types of instruments are envisaged: state of the art *standard* instruments for survey work and back up of 8-m projects, and unique *niche* or experimental instruments for specific projects not feasible on the larger telescopes. The ING should develop near infrared facilities, at least in the J and H bands, but should not try to compete with specialized IR telescopes at longer wavelengths. The Panel believed that it is unlikely to be cost effective to transfer optical work from the ING to United Kingdom Infrared Telescope on Hawaii (UKIRT).

3.2   The Anglo Australian Telescope of the Future

A revised operational model has also been proposed for the Anglo Australian telescope (AAT). While the Australian community needs to maintain the AAT as a general purpose facility, the operational resources will be significantly reduced, possibly to 50% of the present level. A specific future task of the AAT is defined as support of the Gemini telescopes for at least the first decade of Gemini operations. This implies the scheduling of projects at the AAT which require large amounts of observing time which can not be carried out at Gemini. Examples are survey and monitoring programs such as a future major spectroscopic survey following the 2dF galaxy and QSO survey. It was pointed out that digital input catalogues which will become available from the VISTA/VST sky surveys will allow the full scientific exploitation of the VISTA key program by follow up optical spectroscopy over large areas (say 10^4 square degrees), with specific goals such as the search for QSOs at z > 4, radio source identification programs, brown dwarf searches, white dwarf searches, clustering at high redshift, and XMM spectroscopic follow up studies.

3.3   CFHT: 2000 and Beyond

A report on the evolution of the CFHT in the era of 8- to 10-m telescopes has been compiled by Gregory Fahlman and Dennis Crabtree (available on http://www.cfht.hawaii.edu/~crabtree/cfht.html). In line with the conclusions reached by other panels, the future role of the CFHT is seen in the support for the large telescopes but also as a research facility of its own. It has been suggested that large scale surveys, both imaging and spectroscopic, are essential and prerequisites for the scientific programs to be carried out at the large telescopes. These requirements are met with the MEGAPRIME project, which will provide a 1 square degree CCD camera as well as a new wide field corrector to allow for 0.5-arcsec images, and a standardized processing pipeline, TERAPIX, for MEGAPRIME data. A wide field NIR imager, WIRCAM, with a minimum of 4096 x 4096 pixels and approximately a 17' field of view would provide CFHT with a complement to MEGAPRIME. It is noted, however, that the presently available funding will not pay for the

WIRCAM project. Remedies to realize WIRCAM or to gain access at a similar instrument include the acquisition of additional funds via a new partner, the delivery of WIRCAM by a third party in exchange of telescope time, or the exchange of time with another agency who deploys a similar camera on one of their facilities.

3.4   The LaSilla 2000+ Report

The priorities of the ESO user community for the future use of the LaSilla Observatory have been collected through a questionnaire survey by the ESO Users Committee (Nordström 1999). The general consensus by the users is that the medium-term future key role of LaSilla exists in

− optical/IR imaging, precise photometry and polarimetry for survey and monitoring programs and for studies of individual galactic and extragalactic objects, such as the characterization of extra solar planets via photometry of planetary transit light curves,

− optical/IR stellar spectroscopy at high resolution for advanced studies of the hydrodynamics in stellar atmospheres, stellar activity, detailed nucleosynthesis models, and long term coverage for monitoring objects over weeks to study rotating and pulsating stars with longer periods,

− continued access to a competitive (sub)mm telescope until ALMA starts operation

Recommendations for the operation of the LaSilla observatory in the 1999–2006 period were derived from these requirements, by considering the need to reduce the overall operational cost of the LaSilla observatory. Emphasis was put on the fact that nonessential duplication of capabilities on LaSilla must be avoided, and that LaSilla must focus on tasks which are not feasible or not economical with the VLT. It was suggested that all of this will be achieved by specializing the individual telescopes. In particular, the NTT should provide optical and NIR imaging and spectroscopy in preparation for VLT programs. A very large fraction of the 3.6-m telescope should be assigned to planetary system searches (HARPS) and spectroscopy at very high resolution. Two 2-m class telescopes should be dedicated to monitoring. The working group also concluded that LaSilla should not meet the very strong future needs of wide field IR imaging.

4.   OPTICON

OPTICON is a coordination network in European optical and infrared astronomy and brings together the major European national and multinational operators of optical astronomical observatories and archives, and several major research groups and users of the infrastructure. Of direct relevance to OPTICON are optical and infrared telescopes, their instrumentation, observatory infrastructures, data archives, and relevant communications infrastructure. OPTICON provides a forum for the exchange of information between the infrastructure operators, for information feedback from users of the infrastructure, for joint planning of future use and enhancements at its facilities, and for optimization of the scientific access to and exploitation of existing and future new facilities. More detailed information about OPTICON is available via its web page at http://www.astro-opticon.org—not to be confused with the URL http://www.opticon.org, which points to a rock band.

The OPTICON contract partners are the UK Particle Physics and Astronomy Research Council (PPARC), The Chancellor, Masters and Scholars of the University of Cambridge (UCAM/IoA), The Universite Louis Pasteur Centre de Donnees astronomiques de Strasbourg (CDS), the European Space Agency (ESA), the European Southern Observatory (ESO), the Centre National de la Recherche Scientifique, Delegation Regionale Ile de France (Institute d'Astrophysique de Paris IAP), the Institut National des Sciences de l'Universe du Centre National de la Recherche Scientifique (INSU/CNRS), the Instituto de Astrofisica de Canarias (IAC), the Consorzio Nazionale per l'Astronomia e l'Astrofisica (CNAA), the Universiteit Leiden (UL), the Max Planck Institut für Astrophysik (MPA), the Max Planck Institut für Astronomie (MPIA), the Netherlands Research School for Astronomy (NOVA), and the Nordic Optical Telescope Scientific Association (NOTSA).

Among the 14 partners, UCAM/IoA, INSU/CNRS, UL, MPA, MPIA, and NOVA are representatives of the user communities. ESA and ESO are multi national organizations. CNAA, NOVA, and NOTSA are consortia involving several research institutes and are also infrastructure operators. All partners to some extent own, develop, manage, or operate infrastructures, and all are involved in planning and development of future facilities and enhanced access to research and training resources. Each partner is represented in the OPTICON board by national research or funding directors or by research group directors. The OPTICON board meets at least once per year.

As of August 2001, the following OPTICON project programs and working groups have been established:
– Extremely large telescopes

There are a number of concept studies for extremely large telescopes, including ESO's Overwhelmingly Large Telescope (OWL), the California Extremely Large Telescope (CELT), the Swedish 50-m Optical Telescope, the AURA Maximum Aperture Telescope (MAXAT). The OPTICON working group on extremely large telescopes aims to find a consensus for the scientific case of such a project.

− Science requirements for the Virtual Observatory

OPTICON has established three working groups to study the implementation of the Astrophysical Virtual Observatory (AVO). First experience is gained with ASTROVIRTEL, an ESO/ESA/ST ECF initiated program to access some 7 terabytes in archived data from HST, VLT, and some LaSilla telescopes (http://ecf.hq.eso.org/astrovirtel). The working groups focus on scientific requirements, inter operability and GRID technology for AVO. A coordination of AVO with similar proposals in the USA, Canada, and Australia is intended.

− Inter-operability

An OPTICON working group on inter operability studies cost effective tools and standards for improving access to data archives, and for the exchange of data from archives and information services.

− 3D spectroscopy

The EURO 3D (http://www.aip.de./Euro3D) working group joins efforts of European researchers working in the field of 3D spectroscopy. The group has recently submitted a funding proposal to the European Community on 'Promoting 3D Spectroscopy in Europe'. It is presently working on the coordination of developments on 3D data formats for present and future 3D instruments.

− Wide field imaging

The OPTICON working group ASTO-WISE (Astronomical Wide-Field Survey System for Europe) aims to provide a standardized European survey system to facilitate research, data reduction and data mining from future wide field survey cameras. A funding proposal has recently been approved by the EC.

− The operation of medium sized telescopes

See Section 5.

− Technology developments

It is proposed to establish an OPTICON working group on technology developments for future instruments and telescopes, and to study possibilities of funding from the EC and through bilateral industrial programs. Possible issues are AO and MCAO systems, high power lasers, and STJ superconductor tunnel junction detector arrays.

5.   THE FUTURE OPERATION OF MEDIUM-SIZED TELESCOPES IN EUROPE

OPTICON has established a working group to exchange ideas and suggestions for the future operation of medium sized European telescopes. The work plan consists of defining efficient operational modes and means of access for the existing European facilities.

Working group meetings have taken place at the Ecole Nationale Superieur in Lyon in December 2000 and near Calar Alto/Spain in March 2001. For the first time ever, the two meetings brought together the observatory directors and representatives of the funding agencies of the LaSilla Observatory, the AAT, CFHT, UKIRT, TNG, NOT, OHP, Pic du Midi, the planned 2.5-m Greek Aristarchos telescope, the ING and Calar Alto. Valuable input was provided from observers from EAS, PPARC, and the European Northern Observatory on La Palma.

The working group has identified two major goals which should form part of a proposal to the European Community. Firstly, access to the modern research facilities that are operated by the OPTICON partners should be extended to the whole European astronomical community. Second, the working group suggests a significant fraction of telescope time at 1- to 2-m telescopes should be set aside for training and education. A way to realize both goals, possibly within the EC's *Framework Program FP6*, is described below.

5.1   A European Network of Nationally-Operated Telescopes

The future of astronomy lies in the access to large databases and virtual observatories and in cutting edge ground- and space-based telescopes. In Central European countries, astronomers do not have access to these new technologies in general, but also astronomers in some of the present EC states are not always admitted to the most advanced infrastructure which is available to a few national communities only. Researchers throughout Europe require access to major research centers most appropriate to their work, an access that is not restricted by geographical location of the researcher or the research center. Until now, general access to the European national observatories depends largely on the nationality, affiliation, and collaboration, rather than scientific merit of the program of a researcher. While access to Calar Alto and to the ING group from astronomers outside the national communities is possible in principle, the amount of international time granted at both observatories is rather small and does not exceed 5% in general.

The working group suggested that the various national agencies operating telescopes in and outside Europe provide a fraction of the available telescope time to a common European pool. The level at which each agency contributes would be defined and up to the evaluation of each agency. It was also agreed that minimum standards should be established, and that any telescope complying with those standards should be allowed to join. Time allocation for the European network should be arranged via a single panel in order to avoid possible disadvantages which may arise if national panels with possibly different scientific priorities would handle the requests. If it is assumed that some 20% of the total time is made available to a common European pool, funds of the order of 7 million Euros are required to cover travel, data reduction, and operational costs at the various observatories. Interest in joining such a network has been expressed by the ING, Calar Alto, OHP, Pic du Midi, and by European institutions which operate telescopes outside of Europe, among them the Canadian French Hawaii Telescope (CFHT), the United Kingdom Infrared Telescope (UKIRT), the Anglo Australian Telescope (AAT), and ESO's telescopes at the LaSilla Observatory.

The network will allow to integrate Central and Eastern European astronomers into state of the art research, and provide access to the facilities based on scientific merit, opposite to present limitations given by nationalities and the geographical location of a particular establishment. The coordination will foster a culture of cooperation, but also encourage complementary instrumentation at the various facilities which will result in synergism and savings of operational costs. The network will be of very great relevance to countries with relatively little infrastructure investment, and to EC candidate countries. Among the institutes which may benefit greatly from an unrestricted access to all European telescopes are the National Observatory of Athens, the Bulgarian Institute of Astronomy, the Astronomical Institute of the Czech Republic, the Tartu Observatory in Estonia, the Konkoly observatory of the Hungarian Academy of Sciences, the Torun Center for Astronomy and the Nicolaus Copernicus Astronomical Centres in Torun and Warsaw, the University of Ljubljana, the faculty of mathematics and physics of Slovenia, the Vilnius University, to name a few. In a way, the success of such a program has already been proven in the past, where astronomy in Spain has benefited highly from the access to IRAM, ING, and Calar Alto.

5.2 Educational aspects

The working group emphasized that the power of astronomy to motivate young people to study physical sciences should be exploited, and that it is important to train a wide range of young European astronomers for preparation of the use of next generation facilities such as the extremely large telescope. The European Community is recognizing the need to bring research closer to society, to raise the science and technology (S&T) awareness among the younger generation, and to attract the youth to participate in the discussion of S&T. This need suffers from the very strong decline in students studying physical sciences throughout Europe.

Summer schools typically lasting 2 weeks are already in existence which aim to train young astronomy students in the use of telescopes and instruments. A collaboration between Calar Alto, OHP, and Asiago observatory has initiated the NEON (Network of European Observatories in the North) summer schools. The first school was successfully held on Calar Alto in July 2000, and a second school has recently been finished at OHP. Each school has attracted some 15 young Ph.D. students in astronomy. During the course of the schools, the students learned to plan and to efficiently carry out an observing run, followed by data reduction and the summary of the scientific results obtained. The Calar Alto summer school has culminated in the discovery of supernova 200cw in MCG+5-56-007 followed by a scientific publication by the students (Dennefeld et al. 2000). Interest in attending both schools was very high, with about a factor of five more applications received than places available.

Telescope time might also be set aside to attract young students interested in natural sciences, information technologies, and/or high technology instrumentation to careers in research or high tech industry. The inspiration of school children to study physical sciences is already under way in the UK Faulkes project (http://www.faulkes-telescope.com). It provides direct classroom access to two 2-m telescopes situated in the Hawaiian island of Maui and one in Australia, viz. in time zones such that night time observations can be carried out during regular class hours in the UK. Operation of the Faulkes telescopes will be via remote control centers located in the UK, Hawaii and Australia.

5.3   The EC's Framework Programs

The European Community's Framework Programs set out the priorities for the European Community's Research, technological development and demonstration (RTD) activities. The present FP5 program (1998–2002) has a multi-theme structure, with seven specific programs concentrating on the quality of life and management of living resources, user friendly information society, competitive and sustainable growth, energy, environment and sustainable development, confirming the international role of Community research, promotion of innovation and encouragement of small and medium enterprises participation, and improving the human research potential and the socio-economic knowledge base.

Proposals for Council decisions concerning the specific programs implementing the Sixth Framework Program (2002–2006) are presently being discussed. Relevant information may be found on the web page *RTD Beyond 2002* (http://www.cordis.lu/rtd2002). It is presently foreseen FP6 introduces two new concepts: *Networks of Excellence* and *Integrated Projects*. Integrated Projects aim to reinforce European competitiveness through the mobilization of a critical mass of research resources and skills. The objective of networks of excellence is to reinforce European scientific and technological excellence through progressive and lasting integration of research capacities existing in Europe. The size of each network may vary depending on the areas and subjects. The number of participants should not be less than six members. Ideally, a network of excellence is composed of a group of core participants which create a virtual center of excellence. The network partners are from at least three Member States or countries that are associated with the Framework Program. Partners are entities that exist within research centers, universities, or enterprises. As average, the Community financial contribution to a network of excellence, or similar concepts to be introduced in FP6, may represent an amount of several millions Euro.

6.   ACKNOWLEDGMENTS

The present text reproduces parts of a report of the Calar Alto Advisory Committee to the MPIA directorate on the future role of the Calar Alto Observatory, written by R. Bender, R. Gredel, U. Heber, R. Mundt, J. M. Quintana Gonzalez, D. Reimers, and H. Zinnecker.

7.   REFERENCES

Birkle, K.; Elsasser, H.; Neckel, Th.; Schnur, G.; Schwarze, B. 1976 A&A 48, 327.
Dennefeld, M. et al. 2000 IAU Circ. 7457, 4.
Guerrero, M. A.; et al. 1998, New Astronomy Reviews, Volume 42, Issue 6-8, p. 529-532.
Hellemans, A. 2000, Nature 408, 12.
Melchior et al. 1998 A&A 339, 658.
Nordström, B. 1999, The Messenger 98, 57.

Chapter 9

The Roles of Small Telescopes in a Virtual Observatory Environment

S. G. Djorgovski
Palomar Observatory, Caltech
Pasadena, California USA

Abstract: The advent of the Virtual Observatory (VO) concept signals a paradigm shift in the way astronomy will be done in the era of information abundance and ubiquitous networking. Small telescopes will be playing a number of essential roles in this new research environment, probably contributing a major portion of all data taken in astronomy, both as surveying instruments, and as follow-up facilities. In this review we describe the VO concept and its background, and how small telescopes will fit in this emergent way of doing astronomy in the 21st century.

Key words: National Virtual Observatory, surveys, archives, data mining

1.   INTRODUCTION: NEW ASTRONOMY IN THE ERA OF INFORMATION ABUNDANCE

Astronomy, like most sciences, has become immensely data-rich, with data sets measured in many Terabytes, and soon Petabytes. The sky is being surveyed systematically at many wavelengths, with billions of stars, galaxies, quasars, and other objects detected and measured with an unprecedented level of detail. These massive data sets are a new empirical foundation for the astronomy of the 21st century, hopefully leading to a new golden era of discovery.

The steep increase in the volume and complexity of available information is based on the great progress in technology, including digital imaging (the chief data source in astronomy), and, of course, the ways of processing, storing, and accessing information. Most of the scientific measurements and data obtained today are either generated in a digital form, and most

T.D. Oswalt (ed.), The Future of Small Telescopes in the New Millennium, Vol. I, 85–95.

instruments contain digital imaging arrays. Such devices are based on the same technology governed by Moore's law, and are thus growing exponentially in their information-generating ability, with the number of bits in astronomy doubling every 1 to 2 years, while the telescope technology develops more slowly (Szalay & Gray 2001).

Large digital sky surveys and archives now becoming the principal sources of data for astronomy. Most of them are generated using small telescopes. In the past, most astronomical studies dealt with individual objects or small samples thereof (tens to hundreds). Now increasingly the field is being dominated by the analysis of large, uniform sky surveys, sampling millions or billions of sources, and providing tens or hundreds of measured attributes for each of them. This produces a selection of optimal targets to follow up with other telescopes, both large and small, and both ground- and space-based. There is a paradigm shift in observational astronomy, with survey-based science becoming an ever more important way of exploring the universe in a systematic way, and leading to the Virtual Observatory astronomy.

The challenges and opportunities posed by individual massive data sets (surveys and archives) are just a start. The sky is now being surveyed over a full range of wavelengths, giving us, at least in principle, a panchromatic and less biased view of the universe. Each individual sky survey has its own intrinsic limitations in terms of the depth, area coverage, angular resolution, etc., but this can be quantified and at least partially overcome by combining different data sets. Moreover, the universe itself imposes selection effects on what can be seen at any given wavelength: for example, obscuration by dust can hide bursts of star formation and active galactic nuclei from the view in the visible or UV light. Some astrophysical phenomena generate the bulk of their luminosity in a particular region of the spectrum, and may be effectively undetectable or appear inconspicuous as seen at other wavelengths. Federation of surveys and data sets spanning a range of wavelengths (or spatial scales, or time scales, or types of observations, e.g., images and spectra, etc.) can help us discover new knowledge which is present in the data, but not apparent in any of the data sets viewed individually. Joining of massive data sets into even larger and more complex ones thus becomes a scientific imperative.

While the quantity and complexity (and quality) of data continue to grow exponentially, our knowledge and understanding of the world has not kept pace: *we are not making the full use of the information richness available to us.* Transforming vast masses of bits into a refined knowledge and understanding of the universe is a highly complex task. The problems go well beyond technical, and touch the very core of scientific methodology and practice. The great quantitative change in the amount and complexity of available scientific information should lead to a qualitative change in the way we do science.

Old research methodologies, geared to deal with data sets many orders of magnitude smaller and simpler than what we have now, are becoming inadequate. We have to develop new scientific methodologies, which would maximize the utility, effectiveness, and scientific returns of the ever increasing, massive data sets, and enable us to exploit the great opportunities in front of us.

2.   THE VIRTUAL OBSERVATORY CONCEPT

In order to cope with these challenges, the astronomical community started an initiative, the National (and ultimately Global) Virtual Observatory (VO). The VO would federate numerous large digital sky archives, provide the information infrastructure and standards for ingestion of new data and surveys, and develop the computational and analysis tools with which to explore these vast data volumes.

Recognizing the urgent need, the National Academy of Science Astronomy and Astrophysics Survey Committee (McKee, Taylor, *et al* 2001) recommends the establishment of a National Virtual Observatory (NVO) as a first priority in their "small" (i.e., cost <$100M) project category. The early vision of the NVO was presented in the NVO White Paper (2001). In order to provide further community input, to refine the scientific drivers and technological requirements, and draft a roadmap for the NVO, NASA and the NSF formed the NVO Science Definition Team, whose report (NVO SDT 2002) summarises the relevant issues.

This initiative is rapidly growing and gathering support worldwide. Several major international conferences and topical sessions at general conferences have been dedicated to it; see for example, the volumes edited by Brunner *et al.* (2001) and Banday *et al.* (2001), several dedicated sessions at SPIE and ADASS meetings, etc. Several major projects are already under way (see the Web Resources at the end of this Chapter for the relevant links). In what follows, we will assume a trans-national VO concept, although of course any of the regional VO's would play essentially the same roles and collaborate on the global level.

It is important to clarify what the VO is and is not. It is not yet another data archive, or a data center, or a digital library of already produced results, although it will contain links to all of such resources. It rests on the solid and highly successful foundations of the existing data centers, archives, and observatories, but goes well beyond them. It is *a novel type of a scientific research organization* for the Internet era. It has to be geographically distributed, since the data, the expertise, and the computational resources are spread worldwide and are evolving at a rapid pace. It transcends the traditional divisions and agency domains based on the wavelength, observatory- or mission-specific, ground- vs. space-based, etc. The VO will be much more than a depository of massive data sets; it will provide both data-oriented services (archiving and metadata standards, interfaces, protocols, etc.) and data analysis tools and capabilities (survey federation, data mining, statistical analysis, visualization, observations-theory comparisons, etc.). It will have an unprecedented, broad range of users and constituents. *It will be a complete, web-based, distributed research environment for astronomy with massive data sets.* Ultimately, it will become the very fabric of the astronomical practice and become a part of the invisible infrastructure.

The VO will provide new opportunities for scientific discovery that were unimaginable just a few years ago. Entirely new and unexpected scientific results of major significance will emerge from the combined use of the resulting datasets, science that would not be possible from such sets used singly. *The VO will serve as an engine of discovery for astronomy.* At the same time, it will provide great returns to the partnering disciplines, e.g., the applied computer science and information technology (CS/IT), statistics, etc.

The VO concept also represents *a powerful engine for the democratization of science.* Making first-rate resources (the data and the tools for their exploration) available to the entire astronomical community via the Web would engage a much broader pool of talent, including scientists and students from small educational institutions in the U.S. and from countries without access to modern or large telescopes or space observatories, but with strong intellectual traditions. Human intelligence and creativity are now distributed much more widely than the technological resources of the cutting-edge astronomy (or indeed any science). Who knows what great ideas would emerge from this pool? This opening of a scientific opportunity to everyone with a Web access may have a very significant impact in the long run.

Finally, the VO offers unprecedented opportunities for education and public outreach and involvement at all levels: from pre-school to college, including the amateur astronomy community, the media, etc.

3.   FROM THE SURVEY–BASED ASTRONOMY TO THE VO–BASED ASTRONOMY

These developments are already changing the way in which the observational astronomy is done. There is a great increase in the efficiency of use of our most powerful facilities, e.g., large telescopes from the ground or orbiting space observatories. In the past, samples of interesting objects to study have been selected using such expensive facilities themselves; this is demonstrably a poor use of such valuable resources. Now, the most promising targets are selected from large sky surveys, and then followed up with the more expensive resources like the 8- to 10-meter class telescopes or the HST (a small, but highly effective telescope in space). Examples include discoveries of copious numbers of brown dwarfs by the SDSS and 2MASS, and discoveries of large numbers of distant quasars by DPOSS and SDSS, including the most distant objects now known; their follow-up studies are already providing us with valuable new insights into the early universe and structure formation.

The abundance of uniform, high-quality data from large, digital sky surveys is already transforming some fields of astronomy. The nascent field of precision cosmology uses large samples of galaxies and quasars from surveys like SDSS or DPOSS to probe the large-scale structure of the universe and its evolution with unprecedented accuracy. Comparable changes are expected in Galactic astronomy, providing a quantitative description of the structure and dynamics of our own Galaxy at a level of detail never seen before, including a slew of new insights into the stellar astrophysics, the distribution of the dark matter in the universe, etc. These efforts have demonstrated that the old data analysis methodology in astronomy, geared to deal with data sets measured in Megabytes to Gigabytes and samples of objects measured in hundreds and thousands, is not adequate to deal with multi-Terabyte data sets and catalogs of many millions or billions of sources.

The mode of the survey-based astronomy is illustrated schematically in Figure 1. A dedicated (typically small) telescope is generating the data, which are processed and stored in a survey archive. Scientific exploration of this archive through various data mining techniques generates scientific

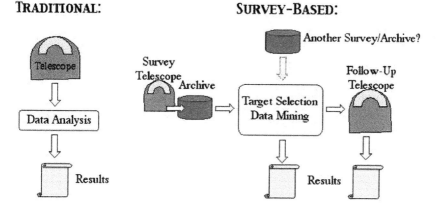

Figure 1. A Schematic Outline of Survey-Based Astronomy

results based on the survey data alone; for example, one can study the large-scale structure in the universe from panoramic imaging alone. In addition, interesting sources are found and followed up spectroscopically or in other ways using other observational facilities (large telescopes, space observatories, etc.); examples may include quasar or brown dwarf candidates, various variable sources, complete samples of galaxies for massive redshift surveys, etc.

Survey-based astronomy is coming of age and producing excellent science. The natural question is, what next? Discoveries made using individual survey, while interesting and valuable, are constrained by the information content of these surveys. There is a lot of added value to be had from data fusion, from federating of the multiple sky surveys to gain a more comprehensive view of the universe. What comes next, beyond the survey science, is the Virtual Observatory science.

Every astronomical observation and every survey covers a portion of the observable parameter space, whose axes include the area coverage, wavelength coverage, limiting flux, etc., and with a limited resolution in angular scales, wavelength, temporal baseline, etc. Each one represents a partial projection of the observable universe, limited by the observational or survey parameters (e.g., pixel sampling, beam size, filters, etc.). Every astronomical data set samples only a small portion of this grand observable parameter space, usually covering only some of the axes and only with a limited dynamical range along each axis. Every survey is also subject to its own selection and measurement limits.

Surveys thus represent hypervolumes in the observable parameter space. Individual sources represent data points (or vectors) in this multidimensional parameter space. So far we have sampled well only a relatively limited set of sub-volumes of this observable parameter space, with a much better coverage along some of the axes than others. Some limits are simply technological or practical, but some are physical, e.g., the quantum noise limits, opacity of the Earth's atmosphere, or Galactic interstellar medium.

Federating multiple surveys that sample different portions of the observable parameter space can provide a much more complex and complete view of the physical universe. The simplest and most traditional, yet very powerful, manifestation of this process is the cross-identification of observations at different wavelengths.

Historical examples abound. The discovery of quasars resulted when some of the first well-localized radio sources were identified in the visible light with what appeared as otherwise non-descript bluish stars − yet they represented a new, spectacular natural phenomenon, and the science of astronomy was changed fundamentally. A similar revelation happened when the sky was mapped for the first time in the far-infrared by the IRAS satellite: it was discovered that the most luminous objects in the nearby universe, identified optically with some irregular galaxies previously considered as mere curiosities, have their powerful energy sources hidden behind the veils of opaque dust. It is now believed that at least a half of all star formation and a large fraction of all AGN in the universe are escaping detection in the standard visible-light techniques due to such obscuration. Cosmic gamma-ray bursts (GRBs) are another spectacular example, where the breakthrough in the understanding of a new astrophysical phenomenon came from insights gained at other wavelengths, including x-ray, optical, and radio: a 30-year cosmic mystery was resolved by combining the data from a range of wavelengths. Essentially all of the x-ray astronomy illuminated the universe in a new light: clusters of galaxies are primarily x-ray emitting objects, and we have gained great insights into a range of variable stars and AGN from the x-ray point of view, with information not available in any other wavelength regime.

These remarkable discoveries represent simple examples of what is possible when data from a range of wavelengths are combined in a systematic way. *Fundamental new phenomena are found, and new insights are gained in the types of objects and processes already known to exist.* We can surely expect even more spectacular discoveries and insights as we start to combine and explore in detail the new generation of massive digital sky surveys, leading us to a better, more complete and deeper understanding of the physical universe.

This synergy of data from many different sources (many of which are generated with small telescopes), illustrated schematically in Figure 2, is a

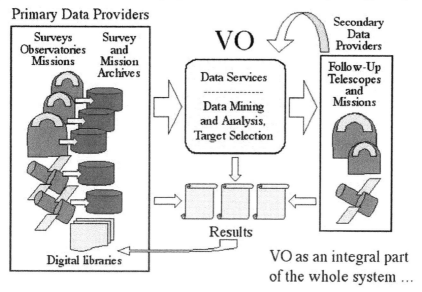

Figure 2. A Schematic Outline of VO-Based Astronomy.

more complex, but also scientifically much more rewarding process, which allows us to fully exploit the information content of the data sets, which we could not do with any of them individually.   Follow-up studies are just as important as in the case of survey-based astronomy, and small telescopes can play a significant role here, especially for the photometric follow-up of variable sources found in synoptic sky surveys. The VO thus plays a role of a unifying and enabling environment for astronomy with a range of facilities and telescopes, from small to large.

A parameter space representation of the available data shows us not only what is known about the universe, but also what are the unexplored areas where genuine new discoveries may be made, and also what is knowable given the technological or physical limits of our measurements (See also Harwit 1975).   *Now for the first time we have the adequate data and technology for such a global, empirical approach to the exploration of the universe.* This has never been possible or done before.

A typical VO data set may be a catalog of $\sim 10^8$–10^9 sources with $\sim 10^2$ measured attributes each, i.e., a set of $\sim 10^9$ data vectors in a ~ 100-dimensional parameter space. Objects of the same physical type would form clusters or sequences in this parameter space. The most populous clusters will be those containing common types of objects, e.g., normal stars or galaxies. But the unusual and rare types (including possible new classes) would form sparse clusters or even be just individual outliers, statistically distinct from the more common ones. Rare objects may be indistinguishable from the more common varieties in some observable parameters, but be separable in other observable axes. This approach has been proven to work and is now used on an industrial scale to find high-redshift quasars and brown dwarfs in SDSS, 2MASS, or DPOSS.

A typical VO project may be a search for the previously unknown types of objects or rare phenomena associated with known types, manifesting themselves through "anomalous" properties, i.e., as "outliers" in some large parameter space. If some type of an interesting object is, for example, one in a million or a billion down to some flux limit, then we need a sample of sources numbering in many millions or billions in order to discover a reasonable sample (or even just individual instances) of such rare species.

Another approach, where we may expect more novelty and surprises, is a systematic exploration of the poorly known portions of the observable parameter space. One such domain is the variability on the sky, especially at faint flux levels, at every wavelength. While a number of variable types of objects are already known, including many types of variable stars, quasars, novae and supernovae, GRBs, etc., we know very little about the time-variable universe in a systematic way (See e.g., Paczynski 2000). There are already some puzzling phenomena found. A few examples include the fast, faint optical transients, which may or may not be associated with distant supernovae, and the mega-flares on otherwise apparently normal stars, which brighten by a factor of a few hundred for a period of hours or days, for as yet unknown reasons.

A new generation of synoptic digital sky surveys, which will image repeatedly large portions of the sky down to very faint limits, will generate the necessary data to start exploring the time domain. A number of NASA-sponsored groups are patrolling the sky with small telescopes, searching for moving objects, and in particular Earth-crossing ("killer" or extinction-causing) asteroids. The same data sets can be also mined for variable objects.

4. CONCLUSION: A SYNERGY OF FACILITIES

As the practice of astronomy undergoes a profound change within the VO paradigm, small telescopes will be among the key constituents of the overall system, fulfilling the essential roles for which larger telescopes are simply not suitable or appropriate. (This is not to say that large telescopes do not play a significant role—they certainly do, especially, but not only for spectroscopic observations.)

First, small telescopes will continue to be the facilities where large sky surveys are generated, especially synoptic surveys, which will open the very exciting new frontier of exploration of the time domain. Second, they can be essential facilities for a rapid, coordinated follow-up of interesting objects and phenomena, again especially in the time domain. These surveys are expected to generate large numbers of viable targets, for which time-critical observations would be essential. Rapid follow-up observations of highly variable or transient sources discovered at any wavelength (from γ-rays to radio) would be essential in understanding and classification of such objects (e.g., a supernova, a GRB afterglow, an OVV AGN, a CV star, ... or something really new). While the VO would provide the analysis environment, the data would have to come from somewhere, and in most cases that may be small telescopes.

Two key issues along this path are telescope dedication and automation. It makes little scientific and operational sense to have universally equipped small telescopes; their scientific and cost effectiveness can be vastly higher if they are dedicated to specific kinds of data gathering, with specific instruments (e.g., a multi-bandpass CCD or IR camera). This is self-evident in the case of survey telescopes, but it is also valid for the follow-up facilities. Likewise, automated operation of such telescopes seems to be a practical necessity. The only exceptions to these requirements may be the telescopes with a strong educational, hands-on component. In many cases it may make sense to form *networks of small telescopes* dedicated for a particular purpose; an example may be the follow-up of GRB observations, e.g., as in the REACT collaboration (http://react.srl.caltech.edu). In any case, the large data volumes and the need for rapid reactions imply that such telescopes must be connected to a proper information infrastructure.

Thus, the future of small telescopes in the VO era seems bright and assured, as the key data providers and as rapid reaction probes of the new and exciting phenomena in the new era of information-rich astronomy.

5.   ACKNOWLEDGMENTS

I wish to thank numerous colleagues and collaborators who contributed to the development of ideas and concepts described here, and in particular R. Brunner, A. Mahabal, T. Prince, R. de Carvalho, R. Gal, S. Odewahn, A. Szalay, and R. Williams. This work was partly supported by grants from the NSF and NASA.

6.   REFERENCES

Banday, A., *et al.* (editors) 2001, *Mining the Sky*, A. ESO Astrophysics Symposia, Berlin: Springer Verlag.

Brunner, R.J., Djorgovski, S.G., & Szalay, A.S. (editors) 2001, *Virtual Observatories of the Future*, ASPCS, Vol. 225.

Djorgovski, S.G., Brunner, R., Mahabal, A., Odewahn, S., de Carvalho, R., Gal, R, Stolorz, P., Granat, R., Curkendall, D., Jacob, J., & Castro, S. 2001, in *Mining the Sky*, Banday, A., *et al.* (eds.) ESO Astrophysics Symposia, Berlin: Springer Verlag, 305.

Fayyad, U., Piatetsky-Shapiro, G., Smyth, P., & Uthurusamy, R (eds.) 1996, *Advances in Knowledge Discovery and Data-Mining*, Boston: AAAI/MIT Press.

Harwit, M. 1975, *QJRAS*, 16, 378

McKee, C., Taylor, J., *et al.* 2000, *Astronomy and Astrophysics in the New Millennium (Decadal Survey)*, National Academy of Science, Astronomy and Astrophysics Survey Committee, Washington D.C.: National Academy Press.

NVO SDT Report 2002, available on line at http://nvosdt.org

NVO White Paper 2001, in: *Virtual Observatories of the Future*, Brunner, R., Djorgovski, S., & Szalay, A. (eds.), ASPCS, 225, 353.

Paczynski, B. 2000, *PASP*, 112, 1281.

Szalay, A.S., & Gray, J. 2001, *Science*, 293, 2037.

7.   SELECTED WEB RESOURCES

The NVO SDT Website: http://www.nvosdt.org
The US NVO ITR Project: http://us-vo.org
The VO Forum: http://voforum.org/
The European AVO Project: http://www.eso.org/projects/avo/
The UK Astrogrid Project: http://www.astrogrid.ac.uk
The author's VO Webpage: http://www.astro.caltech.edu/~george/vo/

8.   SELECTED CURRENT AND FORTHCOMING SKY SURVEYS

Digital Palomar Observatory Sky Survey: http://dposs.caltech.edu
Two-Micron All-Sky Survey: http:// www.ipac.caltech.edu/2mass/
Sloan Digital Sky Survey: http://www.sdss.org
The ESO VST Project: http://www.na.astro.it/vst/
The UK VISTA Project: http://www.vista.ac.uk/

Note: A good set of links on synoptic sky surveys with small telescopes is maintained by B.
Paczynski at: http://www.astro.princeton.edu/

Chapter 10

Developing a Protocol and Implementing a Network for Ubiquitous Use of Telescopes over the Internet: Remote Telescope Mark-up Language — RTML

Carl Pennypacker, Jon Aymon, and Shawn Gordon
University of California, Berkeley, California USA

Robert Denny
DC-3 Dreams, Mesa, Arizona USA

Frederic Hessman
Universitäts-Sternwarte, Goettingen, GERMANY

David Barnaby
Western Kentucky University, Bowling Green, Kentucky USA

Michel Boer
Centre d'Etude Spatiale des Rayonnements, Toulouse FRANCE

Nebosja Duric
University of New Mexico, Albuquerque, New Mexico USA

Toshi Ebisuzaki
Riken Institute, Saitama JAPAN

Gordon Spear
Sonoma State University, Rohnert Park, California USA

T.D. Oswalt (ed.), The Future of Small Telescopes in the New Millennium, Vol. I, 97–112.
© 2003 *Kluwer Academic Publishers. Printed in the Netherlands.*

Vivian Hoette
Yerkes Observatory, Williams Bay, Wisconsin USA

Peter Mack
Astronomical Consultants and Equipment, Tucson, Arizona USA

Abstract: We have drafted a pilot protocol which will enable us to implement a non-homogeneous network of imaging telescopes capable of processing requests for the acquisition and retrieval of simple astronomical images. This protocol is designed to be independent of the specific instrumentation and software that controls the remote and/or robotic telescopes. It embeds traditional astronomical features such as coordinates and exposure times, and allows for prioritized queue scheduling of telescopes while protecting the telescope operating system. The prioritization supports high-stakes interruption of other observations—"Targets of Opportunity" like optical detection of gamma-ray bursts or other transient events. Some generality in this definition and flexibility is desirable, so that a broad variety of objects and observations can be accommodated within this standard. Urgency is needed in the final definition and implementation of this mark-up language, as a variety of satellite, education and science projects can profit by early adoption. A large body of professional and amateur users could use this protocol, and a number of telescope hardware/software companies systems are already working with early versions.

Key words: automated telescopes, robotic telescopes, protocol, satellites

1.   INTRODUCTION

There are a growing number of telescopes available for use over the Internet. Their objectives are very diverse, from scientific to education to public outreach. We refer the reader to the MONET site (http://www.uni-sw.gwdg.de/~hessman/MONET/) for a   list of some of these instruments. The telescopes fall in various categories:

a) Normal telescopes, where an astronomer manually acquires the image, but sends it over the network for a remote usage.

b) Remotely controlled (at least to a certain point), where the users, sometimes located several thousand kilometers away, can interactively control the telescope, acquire images and get them back in real time.

c) Automatic telescopes, working mainly in batch mode. The user, wherever she is located, sends a request for observation. Depending on the combination of hardware, software and local conditions, the telescope schedules the observation (sometimes it can be months after the request), which are then at the disposal of the "observer".

Of course, some mixture of the above operation modes can exist; e.g., a "batch" telescope can have an interactive mode. What creates some unity, both for scientific and other telescopes, is that the person who wants the observation to be acquired can be located far away from the actual telescope, and the actual communication to point the telescope and acquire useful images does not depend on any one system. Developing this idea, we can state that the user sometimes does not want to acquire images from any given telescope, but wants the observing program to be performed whenever some appropriate instrument and time frame are available. This means that both the user and the telescope, or the group of telescopes, need a common network interface, both to send requests and the results of the requests.

In this paper, we present a preliminary draft of such an exchange language, called Robotic Telescope Mark-up Language (RTML). The goal of RTML is to enable communication between user and remote telescopes using a simple but yet powerful interface, based on a widely accepted "extensible mark-up language" (XML), and widely available tools.

RTML is an implementation of XML designed to describe all the information needed to make a simple astronomical observation. The advantage of an XML document over some other format is that it is easily passed as a stream of human-readable strings over the Internet; the syntax is simple and yet rigorously defined; it can be used to describe very complex information; and a large number of software parsing tools exist. In simple terms, XML looks a lot like the HTML used in normal Internet documents (with the mark-up keywords enclosed by "<"'s and ">"'s), but valid documents must adhere strictly to the XML syntax (something which can be validated over the internet). Nowadays, XML is being used to describe everything from spreadsheet contents and public library holdings to NASA spacecraft components. Already a few telescopes and hardware/software companies have implemented RTML capabilities, and a large body of users has the potential to implement it in the near future.

2.   RTML 2.0: AN IMPLEMENTATION OF XML

RTML 2.0 defines XML syntax for the submission of astronomical observing requests to observatories. The goal of RTML 2.0 is to develop a protocol that can work almost immediately with a large number of telescopes. Therefore, it may not embody all of the capability desired for RTML in the long run. For example, no provision for returning results or other operations is provided. In keeping with RTML 2.0's initial goal, only the essential elements are included. Convenience features such as variable units of measure and variable numeric formats are not (yet) supported.

The syntax is defined by a Document Type Definition (DTD) and an XML data-reduced (XDR) schema. The latter is not the W3C standard schema but an earlier type of schema developed by Microsoft. The tools developed by Robert Denny to support RTML 2.0 use the XDR schema for validation, providing data type checks not possible with DTD-based validation. Other implementations may use the DTD for validation. We list in Appendix 1 the current version of RTML (2.0), which contains "tags" (the XML elements contained within the "<"and ">"signs) that are useful for the community of anticipated users, as illustrated in the network diagram below.

2.1   Software Resident on the Telescope Control Computers

2.1.1   RTML Parser

A local parser must take the RTML and extract the validated tags and feed them to the observatory control system. This has been done for Denny's ACP2 and any telescope conforming to the ASCOM standard for telescope and observatory control[1]. In addition, a parser has been written for the Astronomical Consultants and Equipment, Inc. (ACE) automated telescope systems. A wide variety of public-domain XML parsers in practically every language (e.g. Java, Perl, C, C++,..) are available. Although writing a parser takes some time and effort, RTML's rigid definition and relative simplicity makes the job less difficult. In addition, the parser only has to be written once. We anticipate that such parsers for many telescope and computer operating systems will be written and maintained within the GNU context.

2.1.2   Scheduler

An observation scheduler is needed to convert the list of targets from the server into a target list. This operation is independent of the original parsing done at the server, as the main server only checks for appropriate

---

[1] http://ascom-standard.org, http://acp2.dc3.com

instruments, magnitude, and visibility constraints. The local scheduler has the responsibility to account for weather, seeing, partial cloud cover, and other local effects that vary night-to-night. Several schedulers are in use or in development within robotic telescope circles, and the *Hands-On Universe*™ Project intends to provide interested parties with one written in C by Richard Treffers.

2.1.3   Coordinate Refresh Software

Coordinates for moving objects such as planets, asteroids, and the moon must be refreshed in addition to accounting for the non-sidereal motion. We intend to implement IAU or NASA standards for asteroids, planets, and artificial satellites within future renditions of RTML. RTML 2.0 supports the use of a major planet name or a 1-line set of orbital elements (in the *Minor Planet Center* format[2]), in lieu of equatorial coordinates. This permits the observatory to compute real-time positions using well-known algorithms. This has been implemented successfully in ACP2.

2.1.4   Confirming Image Quality (optional)

Ideally, software is needed to confirm that an image is worthy of saving. When an image is taken, an astrometry program should be run to confirm pointing and a photometry program to confirm limiting magnitude. If the image quality is sufficient, as determined by comparing to the limiting magnitude tag in the RTML requests and the pointing is sufficient, the image should be saved and a message sent back to the requestor. This additional service could be provided by the RTML server, though at unnecessarily extra expense, effort, and Internet bandwidth.

2.2   Software Resident on RTML Servers

A main server (eventually migrating to a central set of servers) is anticipated to be used to coordinate and run the network of users. For now, this main server will be hou.lbl.gov.

2.2.1 Sorting Requests and Notifying Requestor

When an observation request arrives at the server, it must be classified according to the capabilities of the network of available telescopes and instruments. This request RTML file is put into a sub-directory at the

---

[2] http://cfa-www.harvard.edu/cfa/ps/info/OpticalObs.html

server, from which the candidate telescopes draw their image requests. Once the file is received, the server replies by sending information on what telescopes are currently available, and what the probable waiting time is, estimated using the average image delivery times over the last month.

When a participating telescope comes on-line; e.g., at the beginning of a night, it retrieves RTML from its sub-directory from the server and signals its intent to perform the observations.

2.2.2 Defining Available Instruments

A database of available instruments and their capabilities must be maintained on the server. An RTML "network crawler" should go out—perhaps once per day—and update this instrument/telescope database. When requests are received, a simple algorithm puts them into the appropriate sub-directory described above.

2.2.3   Scoring Rubric

Ideally, the system should optimize the use of the telescopes on the network by performing sanity checks. Assume a user requests an image or set of images. Using the telescope and instrument database and the priority and waiting lists, the server can determined how closely a particular telescope can fulfill the request(s), and estimate when the observation will be taken. The server can determine which telescopes could fulfill the request and estimate when the observation will be taken

For example, suppose a user wants an image of the nucleus of M82 to search for variability in the J-band, say, once every five days for six months at a variable signal level of 14th magnitude. The scoring rubric would search the telescope and instrument database for a telescope with an infrared camera, determine if the signal to noise is realistic for such an instrument, determine if the telescope is currently available, and make a best estimate of how much of the request could be accommodated by the telescope, both because of the requested exposure times, the achievable signal-to-noise, and the chance of obtaining enough total observing time.

2.3   Software Resident on Remote Observer's Computers

2.3.1 RTML Generation over the Internet

An easy way is needed for filling out RTML forms, with the usual pull-down options for commonly observed objects (and general tags) and the ability to take both time series observations and other types of observations.

Such a form—still in development—now is in Beta 2 stage on the *Hands-On Universe*™ server[3].

2.3.2 Observation Block Builders

A higher level of software—not implemented yet, but clearly desirable—would be a system that can take an entire list of candidate objects and can generate observing "blocks" based on them. Such blocks may consist of observations in several filters, automatic magnitude limit insertion into the RTML form, inclusion of standard stars near the objects, and various types of flat, bias, and dark fields. It would not be hard to generate such a "block generator," as the sub-elements of the blocks (the building blocks??) are in fact RTML elements or documents.

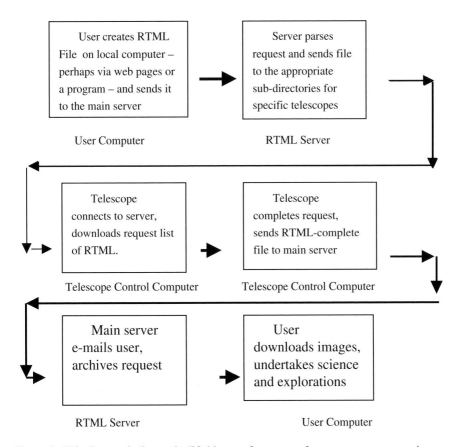

Figure 1. This diagram indicates the life history of a request, from user to server to telescope computer to server to user.

---

[3] http://hou.lbl.gov/RTML

3.   THE HANDS-ON UNIVERSE RTML SERVER

The HOU Server is slowly starting to serve as one source of RTML file request storage and a potential download source of RTML to remote telescopes. Currently, this telescope can generate RTML files from its list of requests, and these RTML files will soon be able to be batch downloaded into the ACP2 system of Denny.   Other download protocols will be developed as the users evolve. That is, if telescope X wants RTML images, it can come to the HOU Server and using standard download syntax, load an observing list from its subdirectory on the server.

3.1   Life History of a Request

Figure 1 outlines the "life history" of a request, through the various computers and elements of this system.

3.2   Network Diagram

The network of users, sources of data, telescopes, and archives is schematically illustrated in Figure 2.

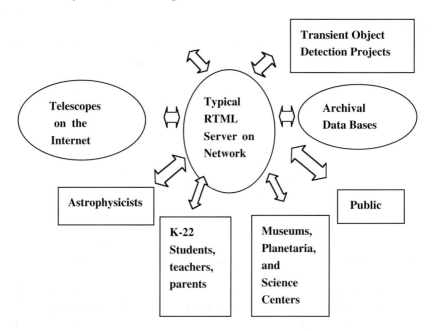

Figure 2. Schematic Diagram of RTML-based system and User and Server Architecture. RTML is the only language used between all of these groups for communications for all object request and retrieval. There can be many RTML Servers on the Network, such as a supernova community RTML server, a GCN server, etc.

3.3   Usage of Local Telescopes

This network intends to enable both full time and intermittent usage of telescopes on the network. A schematic diagram of the use patterns of a typical telescope is shown below in Figure 3.

4.   FUTURE WORK

4.1   Community Feedback

At present, we are implementing this protocol on several telescopes. After successful trial use, we would anticipate a period of community feedback to insure that future modifications (e.g. RTML 3.0) will be useful for a growing community of telescopes and users.

4.2   RTML Complete File

When a request is completed, a new RTML file is completed, but with a different format and DTD or schema than our current RTML. This protocol is yet to be completed, and the existing RTML header with a tag indicating completeness will have to suffice.

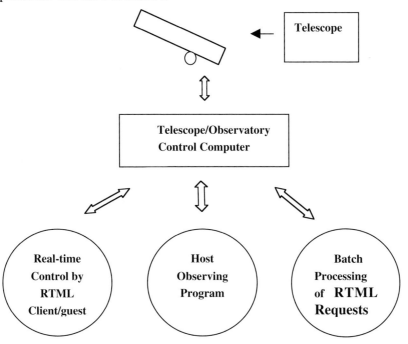

Figure 3. Three suggested modes of operation

4.3   RTML 3.0

After a telescope network based on the exchange of RTML 2.0 requests becomes fully functional, it is clear that we will soon need a more complex description of observation requests. For example, a formal means of distinguishing observing requests from confirmations of those requests and later the documentation of successfully completed requests is needed; more detailed constraints on the desired observing conditions—most importantly seeing—could insure that telescope time is not wasted making observations which the user wouldn't end up analyzing anyway. In addition, observations of binary or pulsating stars require that the user be able to specify the phases during which data is needed (e.g. eclipses). Provisions could be made to permit spectroscopic, polarimetric and even classical photometric measurements. Also, a more flexible use of common astronomical units would make it simpler to define requests. As the networks become more powerful (e.g. networks of networks) and are used by a wider "customer base"—the networks and the users will need a means of "charging" or at least keeping track of the observing time used by various individual users and community of users. Many of these ideas will eventually be implemented in what we call RTML "3.0" as soon as enough experience has been gained in the use of RTML 2.0.

5.   APPENDIX 1: RTML DTD:

An RTML 2.0 document consists of a <Contact> block followed by one or more <Request> blocks. The <Contact> block contains information relating to the person with whom the observatory should communicate regarding the observations, and to whom the results should be returned. A <Request> block contains one or more <Target>s for observation, each of which may contain one or more <Picture>s to take of that <Target>. In addition, the <Request> specifies things like an identification code, the people for whom the observations are being taken, possibly a project name, and other items pertaining to that <Request>.

Figure 4 illustrates the structure of RTML 2.0. The black tags are required and must appear only once. The gray tags are optional (but may appear only once if at all). The red tags are those that can appear more than once and must appear at least once. Finally, the blue tags at a given level are choices; only one of the set may appear at that point. The version = "2.0" is required. Either the DOCTYPE declaration or the x-schema namespace declaration should be given. Either of these may be used by validating XML engines. Any valid file or web URL path is acceptable, as long as it leads to the DTD or XDR schema. To disable validation (not recommended), eliminate both the DOCTYPE and x-schema declarations.

```
<?xml version="1.0" encoding="ISO-8859-1"?>
<!DOCTYPE RTML SYSTEM "http://ASCOM-Standards.org/RTML/RTML-2.0.dtd">
<RTML xmlns="x-schema:http://ASCOM-Standards.org/RTML/RTML-2.0.xdr" version="2.0">

    <Contact> Contact to be used regarding these observations
        <Email>Contact's email address (Internet/SMTP)</Email>
        <User>Contact's name</User>
        <Organization>Contact's organization name</Organization>
    </Contact>
    <Request>
        <ID>Unique request identifier, primarily for HOU database</ID>
        <Username>Username on observatory control system</Username>
        <Observers>Names of people who are requesting observations, e.g., students</Observers>
        <Reason>Reason or objective for requested observations</Reason>
        <Project>Name of project to which observations apply</Project>
        <Schedule> Constraints to be applied to observations
            <Extinct>Maximum extinction through which target may be acquired</Extinct>
            <Airmass>Maximum airmass through which target may be acquired</Airmass>
            <Moon> Constraints relative to Moon
                <Phase>Maximum phase of moon (0..1)</Phase>
                <Distance>Minimum separation, object to moon (degrees)</Distance>
                <Width>Width of avoidance lorentzian (units?)</Width>
            </Moon>
            <TimeRange> Constraints on time, ISO 8601, yyyy-mm-ddThh:mm:ss, UTC
                <Earliest>Must observe after this date/time</Earliest>
                <Latest>Must observe before this date/time</Latest>
            </TimeRange>
            <Priority>1, 2, or 3, 1 is highest</Priority>
        </Schedule>
        <Telescope>Identifies specific telescope to use</Telescope>
        <TimeStamp>Request submission time, ISO 8601, yyyy-mm-ddThh:mm:ss, UTC</TimeStamp>
        <Target
                count="Number of times to repeat entire Target"
                interval="Interval between Target repeats, hours"
        >
            <ID>No specific use defined yet</ID>
            <Name>Target catalog name, suitable for lookup</Name>
            <ObjectType>Any string, will be returned in FITS header as HOUTYPE</ObjectType>
            <Coordinates>
                <RightAscension>J2000 right ascension, degrees</RightAscension>
                <Declination>J2000 declination, degrees</Declination>
                <DateTime>Epoch of coordinates (e.g. minor planet), ISO 8601, yyyy-mm-ddThh:mm:ss, UTC</DateTime>
            </Coordinates>
            <Planet>English name of planet or "Moon", excludes Earth, Sun</Planet>
            <OrbitalElements>IAU MPC "1-line" orbital elements</OrbitalElements>
            <Picture>
                <ExposureTime>Duration of exposure, sec.</ExposureTime>
                <LimitingMagnitude>Depth of exposure, magnitude for point source at 5 sigma</LimitingMagnitude>
                <Filter>Filter name, open for now (see table below)</Filter>
                <RaOffset>RA offset from Target coordinates, degrees</RaOffset>
                <DecOffset>Dec offset from Target coordinates, degrees</DecOffset>
            </Picture>
        </Target>
        <Correction Standard image calibration corrections to be applied
                zero="true | false"
                dark="true | false"
                fixPix="true | false"
                crPix="true | false"
                flat="true | false"
                dome="true | false"
                fringe="true | false"
        />
    </Request>
</RTML>
```

Figure 4. Structure of RTML 2.0.

6.   APPENDIX 2: HOU OBJECT TYPE CODES

Object Type	HOU Code
Unverified cat type	???
Open Cluster	OpC
Globular Cluster	GlC
Diffuse Nebula	DfN
Planetary Nebula	PN
Galaxy	G
Cluster Nebula	GNe
Single Pixel	pix
Star	*
Double or Multiple Star	**
Variable Star	V*
Comet	Cmt
Asteroid	Ast
Major Planet	Plt
Sun	Sun
Moon	MN

7. APPENDIX 3: DIAGRAM OF RTML 2.0

Figure 5 schematically represents the current RTML 2.0 structure.

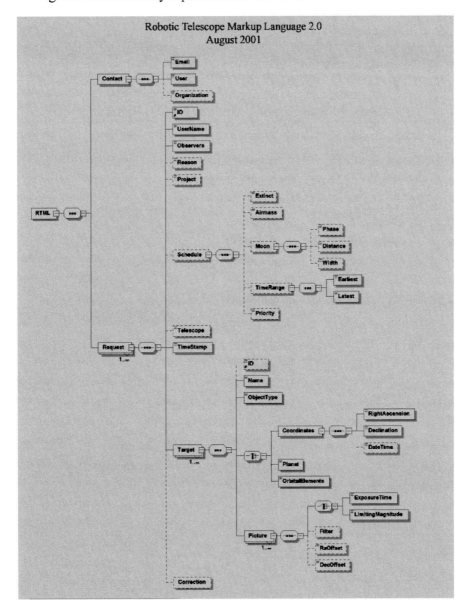

Figure 5. Schematic of RTML 2.0 structure.

8.   APPENDIX 4: EXAMPLES

Table 1 is an example of a minimal RTML request. This specifies a single 180-second exposure of NGC 6705 with no filter and no constraints on the time of observation. It includes a DOCTYPE declaration that references the DTD at the ASCOM web site, permitting validation against the DTD. A file or web URL to any copy of the DTD will suffice. Omitting a DOCTYPE declaration bypasses the DTD validation (not recommended).

Table 1. Example of a Minimal RTML Request

```
<?xml version="1.0" encoding="ISO-8859-1"?>
<!DOCTYPE RTML SYSTEM
        "http://ASCOM-
    Standards.org/RTML/RTML2.0.dtd">
<RTML version="2.0">
    <Contact>
<Email>hsimpson@groening.org</Email>
        <User>Homer Simpson</User>
    </Contact>
    <Request>
        <ID>100-1</ID>
        <Username>homer_s:binky</Username>
        <TimeStamp>2001-07-31T04:02:00</TimeStamp>
        <Target>
            <Name>NGC 6705</Name>
            <ObjectType>OpC</ObjectType>
            <Coordinates>
<RightAscension>282.775</RightAscension>
                <Declination>-6.266667</Declination>
            </Coordinates>
            <Picture>
                <ExposureTime>180</ExposureTime>
            </Picture>
        </Target>
    </Request>
</RTML>
```

Table 2 presents a somewhat more complex RTML request. The RTML tag contains a name space declaration that references the XDR schema at the ASCOM web site, permitting validation against the XDR schema. A file or web URL to any copy of the XDR schema will suffice. To bypass validation, omit the xmlns attribute (not recommended!). This is nowhere nearly as complex as possible, but it does illustrate usage of the repeating elements. The total number of images taken by this RTML is 5 (the first <Request> has <Target count="2">).

Table 2. Example of a More Complex RTML Request

```
<?xml version="1.0" encoding="ISO-8859-1"?>
<RTML                    xmlns="x-schema:http://ASCOM-
Standards.org/RTML/RTML-2.0.xdr"         version="2.0">
        <Contact>
            <Email>rdenny@dc3.com</Email>
    <User>Robert B. Denny</User>
            <Organization>(#663)                       Red
Mountain</Organization>
        </Contact>
        <Request>
            <ID>101</ID>
            <UserName>rdenny:mypasswd</UserName>
            <Observers>Robert B. Denny</Observers>
            <Reason>Test of XRTML</Reason>
            <Schedule>
                <Extinct>0.1</Extinct>
                <Airmass>2.5</Airmass>
                <Priority>1</Priority>
            </Schedule>
            <TimeStamp>2001-07-21T00:00:00</TimeStamp>
            <Target count="2" interval="0.25">
                <Name>IC 986</Name>
                <ObjectType>GlC</ObjectType>
                <Coordinates>

<RightAscension>212.85</RightAscension>
                    <Declination>1.3333</Declination>
                </Coordinates>
                <Picture>
                    <ExposureTime>60</ExposureTime>
                </Picture>
            </Target>
        </Request>
        <Request>
            <ID>102</ID>
            <UserName>rdenny</UserName>
            <Observers>Robert B. Denny</Observers>
            <TimeStamp>2001-07-13T23:46:00</TimeStamp>
            <Target>
                <Name>NGC 5564</Name>
                <ObjectType>GlC</ObjectType>
                <Coordinates>

<RightAscension>215.05</RightAscension>
                    <Declination>7.016667</Declination>
```

Table 2. Example of a More Complex RTML Request, continued...

```
                    </Coordinates>
                    <Picture>
                        <ExposureTime>60</ExposureTime>
                        <Filter>R</Filter>
                    </Picture>
                    <Picture>
                        <ExposureTime>240</ExposureTime>
                        <Filter>B</Filter>
                    </Picture>
                </Target>
                <Target>
                    <Name>NGC 5575</Name>
                    <ObjectType>GlC</ObjectType>
                    <Coordinates>

<RightAscension>215.225</RightAscension>
                    <Declination>6.2000</Declination>
                    </Coordinates>
                    <Picture>
                        <ExposureTime>60</ExposureTime>
                    </Picture>
                </Target>
                <Correction dark="true" flat="true"/>
            </Request>
        </RTML>
```

Chapter 11

Small Telescopes in Astronomy Education

John R. Percy
University of Toronto
Toronto, Ontario CANADA

Abstract: This chapter provides a brief overview of astronomy education, at all levels, and in all settings.   Some of the applications of small telescopes are described.

Key words: astronomy education, small telescopes

1.   INTRODUCTION

This chapter deals with the role of small telescopes in science education, astronomy education in particular.   For a comprehensive discussion of astronomy education, see Hunt (1986), Pasachoff & Percy (1990) and Gouguenheim, McNally & Percy (1998) (proceedings of international conferences on astronomy education), and Percy (1996) for a North American perspective.   Each of these proceedings contains many papers related to small telescopes.   Fraknoi (2000) and Percy (2000a) give shorter but more recent reviews.   See also the web site of International Astronomical Union (IAU) Commission 46 (Astronomy Education and Development)[1].   The emphasis in this chapter will be on education first, and small telescopes second.

Astronomy is useful for many reasons (Percy 2000a): cultural, historical, philosophical, scientific, aesthetic, and pedagogical; these are reasons why astronomy should be included in the curriculum in school and university. (Here, I use the term "university" to refer to all post-secondary educational institutions.)   A small telescope can be used to support most of these goals, though in different ways.   For professional astronomers, there is a completely different reason to include astronomy in the curriculum: to attract and train

---

[1] physics.open.ac.uk/IAU46

T.D. Oswalt (ed.), The Future of Small Telescopes in the New Millennium, Vol. I, 113–123.

the next generation of astronomers. Note that we should interpret the term "astronomer" broadly, to include a wide range of careers in astronomy and related fields; for an enlightening discussion of science careers, see Tobias, Chubin & Aylesworth (1995). I shall not discuss the role of small telescopes in graduate education in astronomy except to point out that (1) small or medium-sized telescopes are excellent for thesis work, since observing time is more readily available, and scheduling is flexible; and (2) as parts of astronomy research are becoming less distinguishable from physics, many astronomy graduate curricula unfortunately no longer require students to take courses in observation and instrumentation.

The potential for small telescopes to support science education is based in part on three developments: the decreasing cost/performance ratio for well-instrumented, computerized small telescopes; the power of the Internet as a channel to control and communicate with small telescopes; and science curriculum reform which emphasizes active learning. Unfortunately there are many forces acting against science curriculum reform (Fraknoi 2000), so much of the potential of small telescopes may never be realized.

2.   SYSTEMS OF EDUCATION, METHODS OF EDUCATION

There are many "systems" of education (Wentzel 1990). In the "European" system, for instance, astronomy is taught in senior high school, as part of a physical science course, and it is taught to *science* students in university. Small telescopes would be useful tools in both of these settings; see McNally (1990) and other papers in the three *international* conference proceedings listed above, for instance.

In the "North American" system, astronomy is taught as part of a general science course in grade 4-9 (age 9-14 years), and to *non-science* students in university. In the US each year, several million students study astronomy at the school level, and at least 200,000 at the university level. Small telescopes do not figure directly or obviously at either level, but they can play a role, both as demonstrations of the tools of astronomy, and through their enrichment value. A much smaller number of *science* students take astronomy courses in university. Some of these will be the astronomers and other scientists of the future. Also, there is increasing emphasis on and support for undergraduate research in North America, and this has had a profound effect on the role of and support for small telescopes.

There are also many *methods* of education. In much of the world, astronomy and the other sciences are taught by memorization and regurgitation of lecture and textbook material. There is no obvious role for telescopes in such a system. Practical or laboratory work can be a useful addition to the "content" material, if it is carefully designed. Practical work is generally equated with "hands-on activities", which are usually thought to be a good thing. I once defined a small telescope as one that I could put my arms around (Percy 1980), so small telescopes would fit the definition of "hands-on". Pennypacker (1992) and Pennypacker & Asbell-Clarke (1996) describe several small-telescope education projects, including the well-known *Hands-On Universe*; the latter reference also lists the advantages of the hands-on approach: it engages students in learning; it encourages them to consider careers in science and technology; it helps them to learn math and science by seeing the relevance of math and science in authentic, genuine settings; it helps them to learn key computational skills. To this list, I would add that it gives them experience and confidence in operating scientific equipment; this may be especially important for students from under-served groups—women, for instance. Pennypacker & Asbell-Clarke (1996) also mention the pitfalls of "hands-on" science, and Hodson (2000) takes up this issue in more detail.

The most effective approach to science education is through "minds-on" processes and activities, which engage the learners in a significant way. This may involve group activities, which model the actual process of scientific research. Education research suggests that social interaction is a significant component of the learning process. A small telescope—local, remote, robotic, or virtual—could be one tool available to such (groups of) learners.

Much is known, from education research, about effective teaching and learning of science. People learn new concepts by building on old ones; students' minds are not "blank slates". People have deeply-rooted misconceptions about scientific topics; in astronomy, these are often based on fundamental concepts such as light and gravity. People have difficulty visualizing three-dimensional concepts, and moving from one frame of reference to another. Concepts must be introduced at the right stage of intellectual development, and in logical order. Teachers at all levels over-estimate what their students learn. Inquiry-based teaching is the most effective. Teaching *more* science should wait on teaching it *better*. Expertise in science does not guarantee expertise in teaching it; pre-service and in-service teacher education is essential at all levels. All education should be subject to research, evaluation, and improvement.

Most university instructors are unaware of this knowledge. They are "amateurs" in the sense that they have received little or no pre-service or in-service education in teaching (in contrast to their many years of preparation in research).

There has been very little research on the role of small telescopes in formal or informal education; in the useful lists of books and papers on education on the web sites of the American Astronomical Society[2] and the Astronomical Society of the Pacific[3], there are no references on this topic. In fact, the education of astronomy majors and other science students in universities is a sorely neglected part of astronomy education discussions—at least in North America.

3.   SMALL TELESCOPES IN GENERAL ASTRONOMY EDUCATION

Students who study astronomy in elementary or junior high school, or in university courses in astronomy for non-science students, may not require small telescopes to "cover the curriculum", or to carry out required practical work, but such telescopes can still play a role. At my university, over 2000 students take astronomy courses each year. Most are non-science students. They have the opportunity to use small telescopes to view astronomical objects. This is an optional, but inspiring, part of the course. The problem, of course, is that "the stars come out at night, the students don't". And it is often cloudy, and always light polluted in my climatically underprivileged city. Nevertheless, a student's first view of our Moon, or Jupiter's moons, or Saturn's rings, through a telescope, is an unforgettable experience.

Most university campuses have a small telescope, often a cumbersome refractor dating from many decades ago.   Some have recently acquired a small telescope through university, government, or private funds; the price of good small telescopes has decreased dramatically in the last two decades. But how to operate the telescopes? We are fortunate, in my department, to have graduate teaching assistants to staff the telescopes.   In many other institutions, the instructors have little or no experience in using a small telescope, and not much time, so the telescope may be under-utilized. Schools usually do not have a telescope, but many amateur astronomers are willing to bring a telescope to class, or to a "star party" in the evening.   The Astronomical Society of the Pacific's *Project ASTRO* facilitates such astronomer-teacher partnerships (Fraknoi & Lalor 2000).

4.   TELESCOPES: LOCAL, REMOTE, ROBOTIC, AND VIRTUAL

This section emphasizes the use of small telescopes in the education of science students, through practical activities and research projects.

---

[2] www.aas.org
[3] www.astrosociety.org

International   Amateur-Professional   Photoelectric   Photometry Communications (ISSN 0886-6961) is a journal that contains a variety of useful papers on this topic. The discussion will be limited to optical telescopes, though small *radio* telescopes have some advantages for undergraduate teaching[4].

Local. Telescopes on university campuses, and other educational settings, have the advantage of being accessible. They can be used by dozens of students with no extra expenditure of time or money. At my university, a 0.4–m telescope is used intensively by astronomy majors in the "Practical Astronomy" course; it is used occasionally and casually by students in other courses; it can also be used by students and the public at our Open House twice a month. For many years, it was also used by undergraduate summer assistants for research, providing them with valuable research experience, as well as a summer job. Our Open House program is organized and conducted by graduate students, who give short non-technical talks, followed by a tour of the campus observatory, and viewing through the telescopes if it is clear. Graduate and undergraduate students can make effective contributions to education and outreach, and should be encouraged to do so.

Remote. Professional research telescopes are routinely located at remote sites, in order to take advantage of sky conditions; the time and cost for the astronomer to travel to the telescope are a necessary and reasonable consequence. To take students to a remote telescope for an education experience might seem prohibitively expensive. In the US, however, there are programs supported by the National Science Foundation, and private foundations, which support undergraduate research. Several consortia of undergraduate institutions have banded together to provide exciting and useful scientific (and social) experiences for undergraduates (DeGioia-Eastwood 1996, Rumstay 1996)). These are undoubtedly important in attracting and training the next generation of scientists.

Robotic. Robotic telescopes have burst onto the scene in the last decade or two. Bond (2000), in a review of a special IAU session on "Astronomy Research Projects for School and University Students" describes several: the *Telescopes in Education* project using the 60-cm telescope on Mount Wilson; the 2-m Liverpool Robotic Telescope to be installed on La Palma with 5 per cent of the time to be earmarked for educational use; the 2-m Faulkes Robotic Telescope to be installed in Hawaii, solely for school use; and the existing Bradford Robotic Telescope which is envisioned as the first of a worldwide network of such telescopes.

Telescopes in the 10- to 40-cm range, for amateur and student use, now come with "go-to" capacity. Combined with the Internet, they can be located anywhere, and controlled from anywhere. It is usually necessary to have a

[4] e.g. www.cassicorp.com, www.haystack.mit.edu

technician on-site, to oversee the telescope, and perhaps to open and close the dome in case of rain. Ideally, "farms" of robotic telescopes could be overseen by a single technician, to keep the costs reasonable, as is done with some professional robotic telescope facilities (Henry 1999).

A single robotic telescope could be under the control of a single user, or a group of users (such as a class or university), or a larger number of unrelated users who access the telescope in a "queue" mode. A problem with one telescope is that it is in daylight (or cloud) for much of the time. This problem can be overcome by having a number of telescopes at different longitudes and latitudes. Another problem is that a single telescope might not seem adequate to serve more than a few students. In many educational settings, however, thousands of students might want the same piece of data (an image of Saturn, the Orion Nebula, or a variable star field, for instance); in this case, a few dozen images, each night, might satisfy the needs of thousands of students. I could envision a national (or international) facility, consisting of a large farm of robotic telescopes, and a *carefully-developed* set of curriculum-connected projects, which would serve tens of thousands of students at any one time.

Virtual. An early example of a "telescope" of this kind was described by Dukes (1990), and a more recent and elegant example—at the UK's Open University—by Norton & Jones (1998). These have many advantages, but they also have the potential danger (in my opinion) of making astronomy into an elaborate video game. We are already at the point where a TV (or WWW) viewer does not know whether they are looking at a real image or a "fake". Perhaps the solution is to combine the virtual telescope with at least one comparable experience with a real local telescope.

5.   ARCHIVAL DATA FROM SMALL TELESCOPES

The volume of astronomical images and measurements from small research telescopes is large, and growing exponentially. Much of the data has never been fully analyzed. This provides students with a gold mine of real data with which to do real science. Image processing is a standard technique among amateur astronomers, for enhancing digital images; software is readily available for using these images for research and education (Dahlman 1994).

In my own field of research—variable stars—online data includes visual measurements by organizations such as the American Association of Variable Star Observers (AAVSO), the epoch photometry of the *Hipparcos* satellite, the "background" measurements of variable stars from gravitational microlensing surveys such as OGLE, and soon MACHO. There is also software for carrying out time series analysis of the data (Percy 2000b). This leads to many opportunities for high school and university students to carry out research projects on variable stars (Percy 2000c). What is needed, for high school and undergraduate students, is *curriculum material and manuals that tie the processes to science and math education standards* (NRC 1996), or to carefully designed expectations of undergraduate courses.

One project that does this is the AAVSO's *Hands-On Astrophysics*. It is based on the premise that astronomers and students can contribute to research by measuring and analyzing variable stars, which yield important information about the nature and evolution of stars in general. Useful measurements can be made by eye, or using binoculars or small telescopes. These measurements can then be sent to databases such as that of the AAVSO for use by scientists and educators worldwide. Data and software are readily available on the WWW (e.g. on the AAVSO web site[5]). For students, variable star observation and analysis can develop and integrate a wide range of science and math skills, while motivating the students through the excitement of doing real science with real data. *Hands-On Astrophysics* includes over 600,000 variable star measurements, as well as software for analysis, charts for making new observations, slides for making practice observations indoors, instructional material including videos, and a comprehensive teachers' and students' manual. Communication and support is available through a web site: hoa.aavso.org. The project can be used at the high school or university level, within a course, or for individual projects.

Projects such as *Hands-On Astrophysics* can be especially successful when carried out by a group of students or amateur astronomers as a cooperative project, since they can build up a communal database of measurements, and compare their individual measurements with those of others. Likewise, they can use archival data for analysis, but add their own new observations to the archive. *Hands-On Astrophysics* is being used in some astronomically developing countries, as a first step in developing an astronomical research capability. To find out more about HOA, or to order the material, contact: AAVSO, 25 Birch Street, Cambridge MA 02138-1205, USA; aavso@aavso.org; web site: hoa.aavso.org.

6. SMALL TELESCOPES AND PUBLIC EDUCATION

Given that most astronomy is financed by taxpayers, the education of the taxpaying public should be a high priority for astronomy—and I refer to more than "glitzy" public relations. It is important for the public to know what

---

[5] www.aavso.org

astronomers do, and why, and to be able to share in the excitement of our science. Unfortunately, the average level of scientific literacy is very low in North America (Fraknoi 2000), and in other countries. But interest in astronomy is high.

There are over 350 public observatories around the world (concentrated in Europe and Japan), according to the *StarGuides* directory maintained by A. Heck at the Observatoire de Strasbourg. These observatories have telescopes that are used for multiple purposes: research by professionals and advanced amateurs, and casual viewing by amateur astronomers, students, and the public. Most public observatories in North America are operated by astronomy clubs, but some are operated by planetariums and science centers. In addition, the visitor centres of major professional observatories are increasingly well funded and used. Both amateur and professional observatories should be encouraged to develop effective, efficient education and outreach programs in partnership with educators; ongoing evaluation and improvement should be part of the program.

It has been my personal experience that to be in an observatory dome, in the dark, with a telescope, is a memorable experience for students and the public. One of the very few education research projects on observatory visitor centers (Burtnyk 2000) supports this view.

7.   AMATEUR ASTRONOMERS, SMALL TELESCOPES, AND EDUCATION

Amateur astronomers are individuals who love astronomy, and cultivate it as a pastime or hobby. There are many "flavors" of amateur astronomers, including armchair amateurs who enjoy reading about astronomy, recreational sky observers who enjoy observing the sky for its own sake, and "serious" amateur astronomers who do astronomy with a high degree of skill, but not for pay (Williams 1988). In North America, the number of individuals in the latter group is comparable with the number of professional astronomers; the total number of amateur astronomers is approximately 40 times larger (Gada, Stern, & Williams 2000). See Percy & Wilson (2000) for a comprehensive discussion of all the ways in which amateur astronomers contribute to research and education. Since amateurs are lifelong learners, grass-roots supporters of astronomy, and citizen participants in astronomy, I think that the encouragement and support of amateur astronomers is a serious responsibility of our profession.

The potential for amateurs to contribute to research has grown dramatically in the last decade, thanks to the availability of good-quality telescopes, CCD cameras, and personal computers at reasonable prices, and thanks to the effectiveness of the Internet as sources of information and communication (Robinson 2000). There is a disturbing trend in many countries, however, called "the graying of the amateurs". Fewer and fewer young people are taking up astronomy as a hobby. Exposure to astronomical images, to the sky, and to small telescopes in school and university may help to change the trend.

Amateur astronomers own the largest share of small telescopes, and they and their telescopes can contribute to education in a dozen ways (Percy 1998). Through star parties and school visits, they can contribute directly and effectively to formal education (Fraknoi & Lalor 2000).

In the AAVSO's *Hands-On Astrophysics* project, processes developed by, and data obtained by amateur astronomers can be used to enhance science and math education. Other observational techniques developed by amateur astronomers (e.g. planetary, solar, occultation, and meteor observing) could be useful for practical work—especially for general courses in observational astronomy (as opposed to more specialized courses which prepare students for future work with large professional telescopes).

8. CONCLUSION

In the context of international astronomy education and development, I recently proposed the concept of "the astronomical community" (Percy 2000d): professional astronomers (if any); other scientists, engineers, and academics with an interest in astronomy; undergraduate and graduate students; educators of all kinds, including schoolteachers and those who educate them; staff of museums and related institutions; science communicators; amateur astronomers of all kinds, at all levels; and interested members of the general public. In this way, astronomers can develop broad-based support for the development and use of their science, and achieve the "critical mass" which is so important at all stages of development. I argued that this concept was relevant to the astronomically developed countries as well. Given the level of scientific (il)literacy in most parts of the world, there are few places that are truly astronomically developed!

The centerpiece of the astronomical community should be a small telescope. Like the traditional public observatory described above, it can be used for many purposes, by most members of the astronomical community. It can be supplemented by a remote/robotic telescope, but the experience of being with and looking through a real telescope should be part of every human being's experience.

9.   REFERENCES:

Bond, P. 2000, *Astronomy & Geophysics*, 41, #6, 30

Burtnyk, K. 2000, *Publ. Astron. Soc. Australia*, 17, 275-281

Dahlman, L. 1994, *Mercury*, Mar.-Apr., 24-27

DeGioia-Eastwood, K. 1996, in *Astronomy Education: Current Developments, Future Coordination*, ed. J.R. Percy, San Francisco: Astronomical Society of the Pacific Conference Series, Volume 79, 190-191

Dukes, R.J. 1990, in *The Teaching of Astronomy*, ed. J.M. Pasachoff & J.R. Percy, Cambridge: Cambridge University Press, 159-170

Fraknoi, A. 2000, in *Amateur-Professional Partnerships in Astronomy*, ed. J.R. Percy & J.B. Wilson, San Francisco: Astronomical Society of the Pacific Conference Series, Volume 220, 243-259

Fraknoi, A. & Lalor, S. 2000, in *Amateur-Professional Partnerships in Astronomy*, ed. J.R. Percy & J.B. Wilson, San Francisco: Astronomical Society of the Pacific Conference Series, Volume 220, 260-263

Gada, A., Stern, A., & Williams, T.R. 2000, in *Amateur-Professional Partnerships in Astronomy*, ed. J.R. Percy & J.B. Wilson, San Francisco: Astronomical Society of the Pacific Conference Series, Volume 220, 14-21

Gouguenheim, L., McNally, D., & Percy, J.R. (editors) 1998, *New Trends in Astronomy Teaching*, Cambridge: Cambridge University Press

Henry, G.W. 1999, *Publ. Astron. Soc. Pacific*, 111, 845-860

Hodson, D. 2000, *OISE Papers in STSE Education*, Volume 1, Toronto: The Ontario Institute for Studies in Education, University of Toronto, 131-143

Hunt, J.J. 1986, *COSMOS: An Educational Challenge*, ESA SP-253

National Research Council 1996, *National Science Education Standards*, Washington: National Academy Press

Norton, A.J. & Jones, M.H. 1998, in *New Trends in Astronomy Teaching*, ed. L. Gouguenheim et al., Cambridge: Cambridge University Press, 100-105

Pasachoff, J.M. & Percy, J.R. (editors) 1990, *The Teaching of Astronomy*, Cambridge: Cambridge University Press

Pennypacker, C. (editor) 1992, *Hands-On Astronomy for Education*, Singapore: World Scientific

Pennypacker, C. & Asbell-Clarke, J. 1996, in *Astronomy Education: Current Developments, Future Coordination*, ed. J.R. Percy, San Francisco: Astronomical Society of the Pacific Conference Series, Volume 79, 61-65

Percy, J.R. (editor) 1996, *Astronomy Education: Current Developments, Future Coordination*, San Francisco: Astronomical Society of the Pacific, Conference Series, Volume 79

Percy, J.R. 1980, *J. Roy. Astron. Soc. Canada*, 74, 334-341

Percy, J.R. 1998, in *New Trends in Teaching Astronomy*, ed. L. Gouguenheim et al., Cambridge: Cambridge University Press, 205-210

Percy, J.R. 2000a, in *Information Handling in Astronomy*, ed. A. Heck, Dordrecht: Kluwer Academic Publishers, 175-185

Percy, J.R. 2000b, in *Amateur-Professional Partnerships in Astronomy*, ed. J.R. Percy & J.B. Wilson, San Francisco: Astronomical Society of the Pacific Conference Series, Volume 220, 212-213

Percy, J.R. 2000c, in *Amateur-Professional Partnerships in Astronomy*, ed. J.R. Percy & J.B. Wilson, San Francisco: Astronomical Society of the Pacific Conference Series, Volume 220, 310-311

Percy, J.R. 2000d, *Teaching of Astronomy in the Asia-Pacific Region*, Bulletin #15, 77-80

Percy, J.R. & Wilson, J.B. (editors) 2000, *Amateur-Professional Partnerships in Astronomy*, San Francisco: Astronomical Society of the Pacific Conference Series, Volume 220

Robinson, L.J. 2000, in *Amateur-Professional Partnerships in Astronomy*, ed. J.R. Percy & J.B. Wilson, San Francisco: Astronomical Society of the Pacific Conference Series, Volume 220, 229-240

Rumstay, K.S. 1996, *IAPPP Communications*, #64, 1-8

Tobias, S., Chubin, D.E., & Aylesworth, K. 1995, *Rethinking Science as a Career*, Tucson AZ: The Research Corporation

Wentzel, D.G. 1990, in *The Teaching of Astronomy*, ed. J.M. Pasachoff & J.R. Percy, Cambridge: Cambridge University Press, 1-6

Williams, T.R. 1988, in *Stargazers: The Contributions of Amateurs to Astronomy*, ed. S. Dunlop & M. Gerbaldi, Berlin: Springer-Verlag, 24-25

Chapter 12

Big Glass on a Silicon Chip: The CLEA Project in the 21st Century

Laurence A. Marschall
Gettysburg College
Gettysburg, Pennsylvania USA

Abstract: In the 1990's, the confluence of personal computing and digital data acquisition made it possible to produce realistic simulations of astronomical instrumentation for classroom use. Since its inception in 1992, Project CLEA (Contemporary Laboratory Experiences in Astronomy) has taken advantage of these developments to introduce a series of PC-based exercises that simulate a wide variety of astronomical research techniques including spectroscopy, CCD imaging, and radio astronomy of pulsars. Project CLEA materials have been widely adopted by astronomy educators because they make it possible to involve students in hands-on investigations that would otherwise be difficult because of time, weather, and financial constraints on real telescopes.

In recent years, Project CLEA software has been developed to take advantage of vastly increased computer power and data base resources as well as the Internet. It is now possible to offer access to tens of millions of objects covering the entire sky, using a variety of simulated observing instruments that span a wide range of wavelengths. As research astronomers plan "virtual observatories" — huge research databases fed by automated survey instruments — the experience provided by well-designed instructional simulations gains additional importance. In the future, it may be only through simulations that most students develop an intuitive understanding of the observational basis of astronomy.

Key words: Education, laboratory exercises, introductory astronomy, computer simulation

1.   INTRODUCTION: CHANGES IN ASTRONOMY AND CHANGES IN ASTRONOMY EDUCATION

In 1900, when "big glass" referred to large refractors like the Yerkes 40-inch, astronomy was undergoing a revolution that was arguably more

T.D. Oswalt (ed.), The Future of Small Telescopes in the New Millennium, Vol. I, 125–138.
© 2003 *Kluwer Academic Publishers. Printed in the Netherlands.*

profound than that going on today. It was clear to most astronomers at the time that astronomy in the 20th Century would be quite different from astronomy as it had been practiced previously. Up to the late 1800's, astronomers could do little more than measure the position and brightness of objects in the heavens, and the analysis of data primarily involved the application of geometry and Newtonian dynamics to problems of time and place.

Textbooks around 1900 reflected the same restricted scope of study (e.g. Young 1902). They were laden with descriptions of coordinate systems and spherical geometry, and included a smattering of material on the telescopic appearance of the planets and the patterns of the constellations. But what the temperature of Saturn was, or what made the stars shine, was purely a matter of speculation. Practical exercises, if there were any, dealt with the reduction of observations, the determination of position and time, or the calculation of orbits.

But things were changing dramatically. The development of photographic and spectroscopic techniques in the mid 1800's had led to the birth of what astronomers of the time (echoing Kepler three centuries earlier), called "the new astronomy" (Langley 1888). Astronomy was becoming astrophysics, and it was now possible to obtain a wide range of physical information about celestial objects by analyzing the stream of photons coming in to a telescope. Far-seeing scientists like George Ellery Hale and Henry Norris Russell were envisioning a new generation of "big glass", huge reflecting telescopes that could collect more light, nourishing a great new age of astrophysical discovery. And so it was, starting with the great reflectors on Mount Wilson, and culminating in the current age of space observatories and 10-meter telescopes.

Judging from what appears in textbooks, the rise of astrophysics in the 20th century brought about a corresponding change in the way astronomy was taught. Russell, Dugan, and Stewart's classic two-volume set (Russell, Dugan, & Stewart 1926), which was the standard college-level introduction to the new astronomy, is notable for its stress on applying physical principles to the understanding of the heavens. By the close of the 20th century, with astrophysics triumphant, introductory astronomy textbooks dealt almost exclusively with planetary, stellar, and cosmological physics, devoting only a few pages to positional astronomy. There is still a simmering debate as to whether introductory astronomy courses should return to the teaching of positional astronomy (Pasachoff 2001; Caton 2001), but the textbooks for such courses no longer exist in print.

On the other hand, astronomy laboratory exercises in the 20th century reflect far less response to the rise of astrophysics than do textbooks. As late as the 1990's, many of the exercises in astronomy laboratory manuals (e.g. Hoff, Kelsey and Neff 1992; Bruck 1990; Ferguson 2001, etc.) involved naked-eye observations of the heavens or the use of small telescopes, not the spectrographs, radio telescopes, and large reflectors that are the principal instruments of modern astronomy. Astrophysical "experiments" in these manuals involved the inspection and measurement of photographic images and spectra, or the manipulation of data from tables of measurements made by others.

As we have noted in previous articles (Marschall 1995, 1998), the phase lag between astronomy textbooks and the astronomy laboratory curriculum was, to a large extent, unavoidable. Despite a general agreement among educators that science students learn best in experiential environments, many of the tools of the astrophysicist are complex and costly, and it would be unreasonable to expect an undergraduate science class to have access to a 10-meter optical telescope or an X-ray imager in space. Moreover, even if such instruments were available, one could hardly ask students to collect the months of observations needed to trace out the light curve of a Mira variable or to assemble the thousands of spectra needed for the large-scale mapping of the galaxies. The time scale required to replicate the important research methods of modern astrophysics is just not compatible with the limited span of an academic calendar.

It is thus no wonder that the creators of astronomy laboratory exercises chose the two options they did. On the one hand, simple observational exercises with the naked eye or small telescopes have the advantage that they give students an immediate acquaintance with the sky. Students like looking at the moon and the planets, even if these observations have little to do with the astrophysical concepts described in most textbooks. On the other hand, astrophysical exercises using canned data have the advantage that they do illustrate how astronomers analyze the results of careful measurements, even if it is not always altogether clear how the data are obtained.

The downside of all of this is that astronomy labs are disconnected from what goes on both in the classroom and at the research observatory. Students often get the impression from these copy-and-calculate exercises that astronomical research is about as exciting as double-entry bookkeeping. Astronomers, as far as they can tell, merely copy numbers from one printed sheet, transfer them to a lab notebook, and grind out graphs and calculations. Lacking a more direct and engaging experience with the instrumentation of 20th century astrophysics, students may confuse the results of research with the phenomena. The pretty pictures of the textbook are taken for the things themselves, and so a student may visualize a pulsar as a spinning ball with magnetic field lines stuck in it like pins in a cushion, not as the regularly-timed series of blips a radio astronomer detects. Students of chemistry recognize early on that a great part of chemistry involves trying to figure out what a nondescript powder is made of, and that a crystal of salt is not just an abstract matrix of balls representing sodium and chlorine atoms. But do astronomy students recognize that astronomers do essentially the same thing? That astronomers spend much of their time trying to distinguish one fuzzy blob of light from another? That astronomers use clever techniques and nifty devices to make discoveries and interpret raw data? That astronomy is far more exciting than double-entry bookkeeping?

2.   PROJECT CLEA: SIMULATING ASTROPHYSICAL RESEARCH ON A PC

2.1   The Origins of Project CLEA

As the 20th century drew to a close, a confluence of technological developments made it possible to resolve some of the difficulties faced by astronomy laboratories in the age of astrophysics. By the late 1980's, computers were becoming ubiquitous in the observatory. They were used for the control of telescopes, for the making of measurements, and for the analysis of data. There was a small but growing body of digital data — images, spectra, photometric measurements, and catalogs — available on tape or disk (and by the mid-1990's, on-line). Personal computers were finding their way into college classrooms and laboratories on a large scale. And a number of programs illustrating physical and astronomical phenomena began to appear in educational circles.

At this time, several astronomy teachers independently developed programs for PC's that were designed to simulate the operation of optical and radio telescopes. Michael Seeds of Franklin and Marshall College produced a photometry program and a spectroscopy program that ran on an Apple II, and Robert Dukes of the College of Charleston developed an ambitious observing simulation, with elements of video gaming, that was designed to work on a Macintosh. John Trasco of the University of Maryland developed similar exercises for early versions of Windows.

These pioneering efforts provided convincing evidence that well-designed computer programs could be used to simulate the process of modern astrophysical research, and it was at this time that Project CLEA got its start. A grant from Gettysburg College in 1991 allowed a colleague, Mike Hayden, and I to develop a first set of computerized exercises aimed at the introductory astronomy laboratory for non-science majors. In 1992 a grant from the National Science Foundation to Gettysburg College enabled us to establish Project CLEA, and to hire a full-time astronomer-programmer, Glenn Snyder, now a co-investigator on the Project. Since that time, with continuing support from the NSF, CLEA has developed 10 computerized exercises in astronomy, which it has distributed widely through the Internet and at workshops around the country. The CLEA materials have met wide acceptance among high school and college teachers, winning several national prizes for educational software, and are currently in use in 50 states and more than 60 nations worldwide.

CLEA exercises attempt to illustrate some of the principal research techniques of modern astronomy, but they vary in complexity and in the elaborateness of the software involved. The first lab developed by CLEA illustrated a simple, classic technique: the observation of Jupiter's moons to determine the mass of the planet using Kepler's laws. Later exercises took advantage of the increase of computer speed and storage capacity to simulate more sophisticated techniques: radio observations of pulsars and spectroscopic investigations of the large-scale structure of galaxies. These developments occurred during the late 1990's, when an explosive growth of computer power in the astronomical research community and a vast increase in the size and connectedness of the Internet opened up exciting new possibilities for education and research. As we enter a new century, it is instructive to review how computer simulations have been applied in astronomy laboratories in recent years, and how research trends in astronomy may affect the course of development of educational simulations in the years to come.

2.2 An Overview of Project CLEA's Astronomy Exercises

Over the past decade, Project CLEA has developed 9 laboratory exercises that are in wide use among astronomy educators. A 10th exercise is available in beta version and is described in the section that follows. Each exercise consists of software, a student workbook, and a technical manual for the use of teachers.

The guiding principles of the Project have been stated in an earlier publication (Marschall 1998), but we review some of the more important guidelines here:

- Exercises are targeted at non-science students in introductory undergraduate courses. But the programs have customizable features that can be set by the teacher to provide a wide range of functionality. CLEA programs are thus usable for upper-class science majors and high school students as well. About 20% of our users, over the years, have been high-school students.

- Most exercises simulate the way astronomers do research at real telescopes; they do not simply present students with drill and practice in concepts.

- Exercises have been only developed where they illustrate techniques that involve costly or complex equipment and inconvenient time scales. If it can be easily observed out of doors (e.g. the phases of the moon), or demonstrated in the laboratory (e.g. the optics of lenses and mirrors), there's no compelling reason to simulate it on a computer.

- Exercises avoid, as much as possible, rote copying and graphing. They emphasize how data is collected and analyzed, and try to engage the student in the actual operation of the simulated device. They do not automatically pop up results.

- Exercises use real data or data produced from "realistic" algorithms.

- Exercises include instructive "real-life" features, such as detector noise, sky background, and weather interruptions, while avoiding complications that are not directly related to the pedagogical goals of the lab, such as the need to adjust detector voltages or to calibrate oscillators.

- Exercises are designed for ease of use and uniformity. Standard Windows grammar is used throughout. Feedback from students and faculty users is incorporated into the software, and significant upgrades are made when warranted.

- Exercises have well-defined learning objectives and goals, which we state in our student manuals. The software and manuals are designed with the cognitive skills and technical sophistication of introductory astronomy students in mind.

The 9 exercises currently available include the following:

- *The Revolution of the Moons of Jupiter*: Students observe the positions of Jupiter's Galilean satellites at regular intervals over several weeks, and determine the period and radius of each satellite's orbit. They use this to determine the mass of Jupiter.

- *Radar Measurement of the Rotation rate of Mercury:* Using a simulated radar telescope, students bounce radio pulses off of Mercury and use the Doppler-broadened echoes to determine the rotation period of the planet and the radial component of its orbital velocity.

- *Astrometry of Asteroids:* Students blink CCD images of asteroids to identify moving objects. They then determine their coordinates using reference stars and the Hubble Guide-Star Catalog, and from this determine the proper motion of the asteroids. A set of images of a Near-Earth asteroid, taken simultaneously from sites separated by several thousand km, is used to determine the parallax of the asteroid, and thus its distance and tangential velocity. A "research" version of this software, *Tools for Astrometry* is also available. It is designed to assist in planning and carrying out observations of asteroids at any observatory equipped with a CCD camera. In additional to blinking and measuring facilities, it can create asteroid ephemerides and charts, and can report results to the IAU Minor Planet Center.

- *The Flow of Energy out of the Sun:* Students can simulate the transport of photons from the center of the sun to the surface. Several exercises explore the physics of random scattering and the formation of spectral lines.

- *Photoelectric Photometry of the Pleiades:* Using a simulated optical telescope equipped with a photoelectric photometer, students can observe stars in an open cluster through several filters, construct an H-R diagram of the cluster, and determine its distance.

- *Classification of Stellar Spectra:* This exercise provides students drill and practice in the identification of stellar spectra. Then, using a simulated telescope with a photon-counting spectrograph, students observe stars, classify their spectra, and determine their distances.

- *Radio Astronomy of Pulsars:* Using a simulated radio telescope, students measure the pulse characteristics of several pulsars from a catalog of over 500. Using the dispersion in the arrival times of pulses at different frequencies, students determine the distance of the pulsars. It is also possible to measure the spin-down rate of the pulsars and to investigate the properties of the radio telescope, such as its beam-width.

- *The Hubble Redshift-Distance Relation:* Using a simulated optical telescope equipped with a photon-counting spectrograph, students observe 5 galaxies in clusters at different distances. They construct a graph of redshift versus distance and use this to determine the Hubble parameter and the expansion age of the universe.

- *The Large-Scale Structure of the Universe:* Using a simulated optical telescope equipped with a photon-counting spectrograph, students, working in groups, observe 10 or 20 galaxies in a limited region of the sky which contains over 200 galaxies. The redshifts of the galaxies serve as a measure of their distance. Combining results with other groups to construct a "wedge diagram" that maps the distribution of galaxies in three dimensions. The data have been selected from the CfA redshift survey (Huchra 1990) and illustrate the overall features of the local distribution of galaxies, including the "Great Wall", and large voids.

Though several of these exercises use similar instrumentation (simulated telescopes, spectrographs, or photometers), they are designed to be independent of one another, so that instructors can adapt them to their particular needs. Manuals and software for each exercise are available on the CLEA website[1] and on CD-ROM by request.

Just enough data is provided with each module to perform the exercise described in the student manual, though it is possible to add data to perform additional exercises. Thus the photometry software can be used to investigate the HR diagrams of clusters of different ages and the spectroscopy software can be used to determine the luminosity function of field stars in the disk of the Milky Way.

The modular form of CLEA software was initially chosen both for pedagogical reasons—to make it more adaptable to different curricula—and for technological reasons—PC's were too slow and had too little storage to handle data covering the entire sky and many different types of objects. But technology has changed, and the 10th CLEA lab represents a departure from the modular design of earlier exercises. It simulates, rather, a multiwavelength observatory with access to the entire sky, and it opens up interesting new educational possibilities. We describe this new development in the section that follows.

2.3   THE SEARCH FOR OBJECT X: THE REALIZATION OF A VIRTUAL EDUCATIONAL OBSERVATORY

The latest exercise under development by CLEA, "The Search For Object X", was designed as a capstone exercise to allow students to apply techniques they learned in other exercises. The concept is simple: students are given the coordinates of an "unknown" and asked use a simulated observatory to find out all they can about it.

Unlike the other exercises, students are not asked specific questions; rather they conduct an open-ended investigation of their own design. The better students will produce more complete reports, and will find that one question leads to another. What kind of an object is it? If it is a star, what are its spectral type, distance, and luminosity? If it is a galaxy, what are its redshift and its distance? If it is a pulsar, what is its period and distance? Some objects require several lines of investigation to make a good identification. Distant galaxies and quasars can look stellar; a spectrum is required to

[1] http://www.gettysburg.edu/academics/physics/clea/CLEAhome.html

distinguish between a star, a quasar, or a galaxy. A stellar object with a G2V spectrum *could* be a star; but if it has moved on a later image, it is most likely an asteroid. This concept is similar to the final experiment in a qualitative chemistry lab, where students are given an unknown and asked to identify it.

The software CLEA has developed for this exercise includes a multi-wavelength observatory with several instruments and data analysis tools, and an extensive database of objects that covers the entire sky. The instruments currently available include three optical telescopes (with apertures of 0.4-m, 1.0-m and 4-m), an IR telescope and a large radio telescope that can be configured as a fully steerable or transit instrument. X-Ray and UV imagers are planned for the future. Instruments include a photometer, spectrometer, and CCD camera for the optical telescopes, and a tunable multi-channel receiver for the radio telescope. Data taken with these instruments can be saved and analyzed using a variety of tools including a digital image display, a spectrum plotter, and a blink comparator and astrometric measuring facility.

The most important innovation in the new software is a large database covering the entire sky. The stellar database, based on the Hubble Guide Star Catalog, provides information on about 20 million stars down to about magnitude 16. Where published photometric and spectral data is available from other sources, it has been incorporated into the catalog, including detailed photometry of 3 or 4 open clusters. Since the purpose of the catalog is educational, we have also generated fictitious color and spectral data for all the remaining objects, based on a plausible model of the spatial and spectral distribution. Data on several thousand galaxies and 500 pulsars has also been incorporated into the catalog, along with an orbital database of 25,000 of the brightest asteroids. We have developed utility software to facilitate adding additional data as it becomes available.

To help instructors of the "Search for Object X" lab, we provide a short list of suggested objects for investigation, but it is clear that the database is far larger than one would ever need to conduct the exercise effectively. Only a vanishingly small percentage of the objects in the database are used by any given classroom; the rest are simply there as background. Or put another way: for the 200,000 students a year that take introductory astronomy classes (Deming & Hufnagel 2001), our database could provide a different object to each student in every class for the next 100 years.

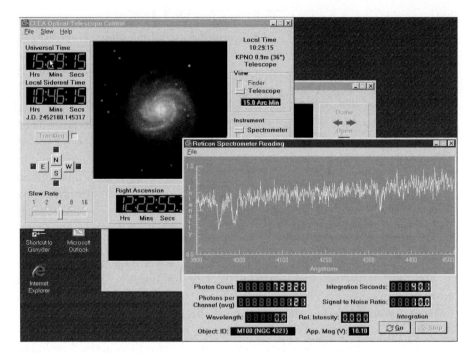

Figure 1. A screen from the Object X exercise

Though it is overkill for one exercise, the large database, coupled with the telescope and instrument simulations, opens up exciting new possibilities. Most of the existing CLEA exercises can be performed using the "Object X" program, and a lot more. Figure 1 shows a typical screen from this exercise.

What we have developed, in effect, is a simulation of an all-sky astrophysical research environment. With the existing software students can simulate the discovery of asteroids by taking repeated images of selected areas of the sky. They can investigate the H-R diagrams of clusters of different ages. They can collect statistics on the distribution of spectral types or hunt for distant quasars. Though it is far from a complete accounting of everything we know about the sky, the database is already so large that it approximates the situation at a real observatory. There is more information implicitly contained in the database than in our explicit design of the exercise, and just like a real observatory there are "so many stars, so little time".

In the past developing a new CLEA exercise involved programming new software and assembling new data. In the future, using the Object X software, the development process will most likely involve the identification of suitable objects already in the database and production of student manuals and teacher's guides to support the particular investigation. For educational purposes, at least, we have put the sky and observatory on a silicon chip.

2.4 NEXT STEPS: OLEO, AN ON-LINE EDUCATIONAL OBSERVATORY

While the development of the Object X lab was continuing, Project CLEA received funding for the next step in its development. CLEA software currently consists of individual programs that run in a MS-Windows environment. Over the next three years (2002 through 2004), Project CLEA intends to go on-line, developing an On-Line Educational Observatory (OLEO) based on the model of the Object X software. The details of the system have yet to be determined, but students using OLEO will be able to access a large all-sky database over the web using software (possibly browser-based) that resembles the operation of the current CLEA instrumentation.

This is a logical extension of current programs. The earliest CLEA labs fit easily on a single floppy, but database for the Object X lab is already so large that it can barely fit on a single CD-ROM, and the download time to most users over the web is prohibitively long. While we could distribute CLEA software on DVD's, a far more workable solution seems to be maintaining the database on a central server, which would send, on demand, the much more modest amounts of data required by each student user. Moreover additions and corrections to the database can be easily made on a central server, forgoing the need for downloadable updates. With the development of web-based observing software and a database that can be expanded and modified, future development of CLEA exercises will center primarily on developing manuals and lesson-plans to make use of the extensive resources of OLEO.

3. CLOSING COMMENTS: BIG GLASS, BIG DATABASES, AND THE ROLE OF OBSERVATIONAL SIMULATIONS IN ASTRONOMY EDUCATION

It is notable, and probably not entirely coincidental, that research in professional astronomy has been developing along lines parallel to our educational effort. Large research databases, fed by surveys such as the Sloan Digital Sky Survey, the 2dF galaxy survey, and the 2Mass infrared survey, have led astronomers to consider new ways of organizing the data that is gushing into their computers. There is currently serious consideration of a National Virtual Observatory (NVO), an on-line archive of data organized so that astronomers could use it as an instrument of astronomical discovery (Szalay & Gray 2001).

A central feature of both the Object X lab and Project OLEO are large databases organized so that students can to use them to conduct realistic simulations of astronomical observations. The forces that led to their development are similar to those that are leading, albeit on a much larger scale, to the NVO: huge increases in computer power, the ease and speed with which digital data can be collected and stored, the ballooning of data storage capacity, and the growth of the internet.

How might these developments affect educational simulations in the future? In the past, one of the principal objectives of Project CLEA was to simulate the techniques of research astrophysics in a controlled educational environment. The model it used was that of the individual astronomer in control of a telescope, able to point it to individual objects of interest.

But research astronomers of the future may not, in fact, have such an intimate role in collecting data. Already many observatories, especially those in space, are run by a technical staff which collects data requested in a proposal and makes it available, in semi-processed form, on a digital archive. Already many astronomers are becoming accustomed to just downloading files from an archival server, never having seen the "front end" of the device that produced the spectrum or took the image they are using.

In the not-so-distant future, most astrophysical research may be done on the computer, with astronomers retrieving data collected robotically and massaging it through a variety of processing and visualization tools. Thus, without explicitly intending to do so, Project CLEA software may indeed be simulating the research environment of the future. As the NVO database becomes a reality, in fact, it may be possible to make use of its archives to make the CLEA educational database more realistic. Students may, in time, have access to the same data that research astronomers do. The expensive and complex tools that observe the heavens may be as physically remote from astrophysical researchers in the future as they presently are to undergraduate astronomy students.

Nevertheless Project CLEA aims to continue developing simulation software along the model of an individual astronomer in control of a real telescope. Even though it may draw on a large database of remotely collected data, it will remain a principal feature of CLEA materials that they engage students in the process by which data is collected using simulated telescopes and observing instruments. Were this not the case, we might just as well return to the days in which students were handed a photograph and asked to apply a measuring ruler to it.

It is not difficult to argue, as we did in earlier sections of this paper, that such individual engagement is essential to effective learning, at least at in introductory classes for non-science majors. But in the future, realistic simulations along these lines may also be of value in classes for astronomy and physics majors.

Even—or especially—for research astronomers, remote observing comes with a price. Astronomers profit by understanding the capabilities and limitations of their equipment, by pushing the envelope just enough to advance the frontiers of astrophysical research, and they cannot do so if all they see is the data deposited in a computerized archive. Radio astronomer Felix Lockman argued this case most cogently in an essay "Can Remote Observing be Good Observing?" (Lockman 1993). Astronomers of the future may, he wrote, "become intellectually weakened by distance, by separation, transformed into consumers of data, not experimental scientists or producers of it."

It is possible that CLEA's simulated telescopes, or some future software developed along similar lines, may be one of the principal ways in which upper-class astronomy students may learn how to carry out astrophysical observations in an era in which telescopes are too big or too remote to be operated by the individual doing the research. Just as professional pilots train first on simulators, astronomers may serve part of an apprenticeship at a simulated observatory. In the age of big glass, telescopes on little chips of silicon may, paradoxically, be even more useful in astronomy education than they are today.

4. ACKNOWLEDGEMENTS

Project CLEA would not be a success were it not for my industrious co-investigators over the last decade, Glenn Snyder, Dick Cooper, Mike Hayden, Rhonda Good, Mia Luehrmann, and Helenmarie Hofman. Mia and Helenmarie first promoted the idea of the Object X lab in the summer of 1994, a long time before it became technologically possible to do it on a PC. Thanks also to our student assistants over the years, especially Shawn Baker, Michelle Vojtush, Akbar Rizvi, Julia Lynch, Brin Finnigan, Nazir Tyrewalla, Laura Jones, and Jennifer Frohnapfel. Thanks to our evaluators, Marcus Lieberman and Michael Chabin, to the participants in our summer workshops, and to the many astronomers, teachers, and amateurs around the world who have offered criticism and encouragement. Project CLEA has been supported by grants from the National Science Foundation and Gettysburg College.

5.   REFERENCES

Bruck, M.T. 1990, Exercises in Practical Astronomy Using Photographs (Bristol, UK: Adam Hilger)

Caton, D. 2001, The Physics Teacher, 39, 382

Deming, G., & Hufnagel, B. 2001, The Physics Teacher, 39, 268

Ferguson, Dale. 2001, Introductory Astronomy Exercises, (2nd edition, Pacific Grove, CA: Brooks-Cole)

Hoff, D.B., Kelsey, L.J, & Neff, J.S. 1992, Activities in Astronomy (3rd Edition, Dubuque, IA: Kendall-Hunt Publishing Co.)

Huchra, J.P, CfA Redshift Catalogue, (Harvard-Smithsonian Center for Astrophysics, 1990.) Available from the Astronomical Data Center (http://adc.gsfc.nasa.gov)

Langley, S.P. 1888, The New Astronomy (Boston: Ticknor)

Lockman, F. 1993, in Observing at a Distance, eds. D.T. Emerson and R.G. Clowes (Singapore: World Scientific)

Marschall, L.A.. 1995, Sky and Telescope, 90, 92

Marschall, L.A.. 1998, in New Trends in Astronomy Teaching, eds. L. Gouguenheim,, D. McNally, & J.R. Percy., (Cambridge, UK: Cambridge University Press), 79-87

Pasachoff, J. M. 2001, The Physics Teacher, 39, 381

Russell, H.N., Dugan, R.S., & Stewart, J.Q. 1927, Astronomy; a revision of Young's Manual of Astronomy (Boston: Ginn and Company)

Szalay, A, & Gray, J. 2001, Science, 293, 2037

Young, C.A.. 1902, Manual of Astronomy; a textbook (Boston: Ginn and Company)

Chapter 13

The International Small Telescope Cooperative (ISTeC)

Gary D. Henson
East Tennessee State University
Johnson City, Tennessee USA

Abstract: Although the emphasis in mainstream astronomy has moved increasingly toward larger and space-based telescopes, it is the small to intermediate size telescopes (in the 0.4- to 2.0-m range) that have been the backbone upon which contemporary astronomy has been built. Numerous institutions and individuals with such small aperture telescopes exist throughout the world. Such facilities tend to be largely undersubscribed yet they are ideal for supporting areas of research that cannot or will not be scheduled on large telescopes. The International Small Telescope Cooperative (ISTeC) was established to bring to the attention of the general astronomical community the distribution and availability of such small telescope facilities and their research capabilities. The cooperative functions mainly through an Internet website that contains an information listing of the facilities that have volunteered to become a member of the cooperative. The purpose of this listing is to provide research astronomers worldwide with potential sources of data on a larger scale and in a more time efficient manner than is currently available using large national facilities alone. Looking toward the future, advances in telescope automation, detector technology, and worldwide communications will provide the opportunity to make even the smallest of telescopes a productive research instrument. It is the intent of ISTeC to preserve and advance the role of small telescopes in astronomical research by fostering both national and international research collaborations via the Internet.

Key words: ISTeC, NASTeC, cooperatives, collaborations, database

1.   HISTORY OF ISTEC

The International Small Telescope Cooperative (ISTeC, pronounced "ice-tech") was originally founded as the North American Small Telescope Cooperative (NASTeC), by Dr. Jason Cardelli in 1992. NASTeC existed

T.D. Oswalt (ed.), The Future of Small Telescopes in the New Millennium, Vol. I, 139–145.
© 2003 *Kluwer Academic Publishers. Printed in the Netherlands.*

initially through an anonymous FTP site that listed the equipment, capabilities, and research interests of participating small telescope facilities. Participation in the listing was and continues to be voluntary with no cost or obligation to cooperative members other than a desire to contribute to astronomy research and education. The FTP site quickly evolved into a website so the database of facilities was easily available to anyone with Internet access.

In May 1996, Dr. Cardelli died suddenly and tragically leaving the future of NASTeC in doubt. A call went out to the users of small telescopes to identify a new coordinator for the cooperative. The call was answered by the Southeastern Association for Research in Astronomy (SARA), a consortium of universities in the southeastern U.S., which operates a 0.9 meter telescope on Kitt Peak, Arizona. SARA took over the responsibility of serving as an institutional home for the cooperative's webpages in the fall of 1996. Heather Preston and then Fred Ringwald managed the webpages for several years. During this time the cooperative grew to include facilities outside the North American continent. It was renamed the International Small Telescope Cooperative in 1999. Currently, this author, a faculty member at one of the consortium's founding institutions, is serving as the ISTeC co-ordinator for the webpages and is responsible for collecting and updating the information they contain.

2.   PURPOSE OF ISTEC

The primary goal of ISTeC is to bring to the attention of the general astronomical community the distribution and availability of small-to-intermediate size research telescopes and their associated equipment through a voluntary information listing. This listing of telescopes serves to help observational astronomers obtain multi-mode, multi-wavelength, or target-of-opportunity observations, in addition to providing the incentive to develop new and broader scientific collaborations using small telescopes. The existence of the cooperative also provides more opportunities for graduate and undergraduate students to become involved in a broad spectrum of astronomical research.

ISTeC grew partly out of the concern that small telescopes were being forgotten, ignored, or deemed unworthy of significant contributions to science. In the past few decades, the emphasis in mainstream astronomy has moved increasingly toward larger telescopes and the numerous scientific benefits they bring. While the need for larger telescopes is without dispute, the reality of limited financial resources has resulted in a general decline in the status of small-to-intermediate size telescopes (roughly those in the 0.4-

to 2.0-m class). As a consequence, many institutions with such telescopes have cut back or stopped instrument development and in some cases have allowed their facilities to fall into disrepair and disuse. The potential loss of small telescope facilities represents a serious loss to contemporary astronomical research. The limited number of large telescopes for the foreseeable future means that the majority of the available time will be dedicated to the type of science for which such telescopes are best suited (e.g. faint target photometry, imaging, and spectroscopy). Such limited time resources means that observing needs cannot be met in many areas of research involving relatively bright targets that do not necessarily require large telescopes. Thus, with the loss of small telescopes, such science may simply not be done.

There are numerous ways in which small telescopes can continue to contribute to astronomy. Although dependent on the individual institution, small-to-intermediate size telescopes tend to be undersubscribed in comparison to larger telescopes. Thus, in addition to supporting areas of research that cannot or will not be scheduled on large telescopes, the availability of small telescopes is particularly important in time-critical or "target-of-opportunity" scenarios. Small telescope facilities also provide a valuable resource for obtaining data in collaboration with other ground-based programs or programs centered on observations using spacecraft like the Extreme Ultraviolet Explorer (EUVE), the Infrared Space Observatory (ISO) and the Hubble Space Telescope (HST). In the past, such interaction has proven to be extremely useful in creating and fostering scientific collaborations between both individuals and institutions. Additionally, small telescopes are important tools in the education of students and collaborations with outside groups provide a potential means of enhancing student research training. Small telescopes remain an invaluable resource for astronomy that should be recognized and utilized by the astronomical community.

The major objective of ISTeC is the creation and distribution of an information listing of small-to-intermediate telescopes and their associated instrumentation available throughout the world. The principal purpose of the listing is to act as a central registry for small telescope facilities where astronomers using and/or needing small telescopes can find each other. Many observational projects could effectively use these instruments, but more often than not astronomers involved in such projects are unaware of potential collaborating facilities. ISTeC provides research astronomers with potential sources of observational data on a larger scale and in a more time efficient manner than would normally be available through use of a single large telescope facility. It is the intent of ISTeC to preserve and advance the role of small telescopes in astronomical research by fostering the exchange of observations and the development of collaborations via the information listing at an Internet website.

The information obtainable through ISTeC contributes to both research and education in astronomy in many ways. In the area of research, it is the purpose of ISTeC to provide: (1) a means of obtaining multi-mode data (e.g. photometry, spectroscopy, imaging, etc.) in a time-efficient fashion, (2) a means to exploit multi-wavelength science by linking ground based capabilities with data obtained over a wide range of energies (e.g. ultraviolet, infrared, radio, etc.), (3) a means to exploit multi-wavelength and/or multi-mode science in time-critical or "target of opportunity" scenarios (e.g. supernovae, novae, variable stars, etc.), and (4) a means by which potential scientific collaborations can be established and expanded, especially for individuals and institutions that are not in the mainstream of astronomical research. In the area of education, it is the purpose of ISTeC to provide: (1) an opportunity for institutions with small astronomy programs to become more involved with mainstream astronomical research, (2) an opportunity for graduate as well as undergraduate students to become directly involved in a broad spectrum of research, (3) justification and impetus for institutions associated with astronomical research and education on all levels to more actively maintain and expand their telescope and instrument capabilities, and (4) a means to involve serious amateur astronomers in mainstream research which can foster better understanding and support from the general public.

3.   ISTEC WEBSITE

ISTeC currently consists of over 80 institutions or facilities operating over 100 telescopes in over half the US states, Canada, Mexico, western and eastern Europe, South Africa, and New Zealand. The current address for the ISTeC webpages can be found at the end of this article. Given the dynamics of web servers and site names, a permanent link to these pages can always be found through the main pages of the host organization, SARA (this web address is also listed at the end of this article). The home page provides a brief description of ISTeC including its history, purpose, and organization. There are also links to more detailed descriptions about the background and goals of ISTeC and to some small telescope networks (GNAT, WET, etc.) already in place.

Most important on the ISTeC homepage are the links to a web-based enrollment form and the information listing of telescope facilities. The enrollment form is entirely completed and submitted on the web. Although any observatory can fill in and submit the form in just a few minutes, the information is checked and the validity of the facility verified by the co-ordinator before the data are entered into the master listing. In addition to entering basic information about the location, telescopes, instruments, and a contact person, there are sections for comments and requirements. These sections generally contain such information as how much advance notice is required, weather statistics, the member institutions schedule load, observing experience of the staff, etc. At the present time, the information listing consists of two parts (*this is in the process of change, see "Future" section*). The first part contains a "quick reference guide" (one line per telescope per institution) giving general information such as location, telescope size, and general capabilities (e.g. photometry, spectroscopy, imaging, etc.). The "quick reference guide" can be sorted alphabetically, by telescope aperture, or by longitude of the observatory. The "main listing" is a more detailed summary which includes all information submitted on the enrollment form. This includes details of the available equipment, site characteristics (e.g. number of clear nights, etc.), contact persons, and specific requirements for users.

Participation in ISTeC is purely voluntary and individual institutions must agree to appear in the listing. More importantly, participation in the listing is strictly non-binding. A participating ISTeC institution is neither required nor obligated in any way to honor a request from any individual or group. Furthermore, the guidelines regarding all aspects of when and how the data are to be obtained or processed are determined by the participating institution. Potential users simply access the database to find the facilities that could possibly contribute to their project. The user then directly contacts the contact person listed for the facility and the collaboration proceeds from there. Potential users are directed to look carefully at comments and requirements for the individual institutions in addition to their observing capabilities.

4.   FUTURE OF ISTEC

It is clear that small telescopes remain a valuable resource for astronomy that must be preserved. Efforts in the US towards this end have begun in recent years through a workshop, the First Annual Lowell Observatory Fall Workshop: The Role of Small Telescopes in Modern Astronomy, which took place in October 1996[1] and through two special one-day sessions at the January 1997 meeting of the American Astronomical Society in Toronto, Canada, and at the June 1997 meeting of the American Astronomical Society in Winston-Salem, North Carolina[2]. More conferences dealing with the role small telescopes should play in the future of astronomy have begun to be held internationally. These meetings emphasized the importance of small

---

[1] See abstracts and comments at http://www.noao.edu/aura/stma/small_telescopes.html
[2] See abstracts at http://www.aas.org/meetings/past_meetings.html

telescopes in providing the essential training in observational astronomy for students at all levels, in providing ground-based observational support with the time and geographic constraints often necessary for multi-wavelength and spacecraft observations, in providing nearly unlimited telescope time for long-term and survey projects, and in serving as testbeds for new instrumentation and modes of operation.

Advances in telescope automation and detector technology have dramatically improved the performance and efficiency of even the smallest of telescopes. In addition, new designs for small telescopes are bringing the cost significantly down, even for 1-m class telescopes. The cost for standard instruments such as CCD cameras has also been decreasing. Improvements in the reliability and bandwidth of the Internet have made remote operation of telescopes (in real-time) and the transfer of large amounts of data routine for locations all over the world. Just as importantly, such advances and improvements will increase the capabilities and productivity of telescopes owned by amateur astronomers. Many amateurs have long contributed to published research and ISTeC can open the door for many others in support of professional-amateur collaborations. Thus, it is likely the number of research-capable small telescopes will actually increase in the future, enhancing the importance of coordination efforts.

ISTeC is working to alleviate one of the main problems limiting the utilization of small telescopes around the world, their relative isolation. The original mission of ISTeC was a passive one. Although the information listing has been and will remain a critical resource, ISTeC will begin to take a more active role in seeking out new members, encouraging communication and exchanges of information between members, and in announcing opportunities for participation in research programs.

Plans are already in place to create a listserver for ISTeC members under the control of the co-ordinator. This listserver will facilitate communication between the members and allow the efficient dissemination of any "announcement of opportunity" from potential users needing observations to be made. Also, acquisition of the ISTeC database will be made more user-friendly and searchable. Potential users will be able to more easily identify and contact the best matches to their observing needs. Finally, ISTeC will more aggressively seek new members and attempt to become more visible to the astronomical community through periodic announcements, updates, or newsletter articles.

ISTeC will remain a "volunteer" organization and the volunteer spirit is alive and well within the small telescope community. These volunteers are ready, willing, and able. It's time to put them to work!

5.   ACKNOWLEDGEMENTS

ISTeC would not exist if not for the passion and effort of Jason Cardelli. The astronomical community was saddened by his untimely death and the continued existence of ISTeC recognizes in some way the contributions of Jason to his chosen profession.   Much of what is written here has been extracted from the information and descriptions found on the ISTeC webpages.   Both Heather Preston and Fred Ringwald, although no longer associated with the SARA consortium, made important contributions in the development of these webpages and in the growth of the membership in ISTeC. Their efforts are greatly appreciated. Finally, the SARA consortium should be commended for its role in helping to represent the interests of astronomers who need small telescopes to conduct their research.

To find ISTeC on the Internet go to http://astro.fit.edu/istec, or visit the SARA homepage at www.saraobservatory.org and follow the link to ISTeC.

Chapter 14

Development, Construction and Plans for Small Telescopes in Japan

Takehiko Kuroda and Fumiaki Tsumuraya
Nishi-Harima Astronomical Observatory
Sayo, Hyogo JAPAN

Abstract: The establishment of the Okayama Astrophysical Observatory (OAO) has initiated modern observational astronomy in Japan.   Almost all of their observatories are established and operated by local governments for the public service.   The telescopes are getting larger every year and the instruments are in the process of upgrading.   Activities of these public observatories are attracting much public attention to the astronomy community.

Key words: public observatories, national observatories, university observatories

1.   INTRODUCTION

Modern observational astronomy in Japan has been initiated by the establishment of the Okayama Astrophysical Observatory (OAO), which belongs to the National Astronomical Observatory of Japan (NAOJ). However the aperture of the main telescope had only been a 1.88-m reflector and was far behind the trend in the world.   The Subaru telescope finally completed test observations in 1999, and began observations in 2000 for astronomers in Japan and other countries.   This telescope has an 8.2-m aperture, which is world-class, but it is located in Hawaii.

On the other hand, the number of 0.60-m to 1.5-m class telescopes has increased in the past twenty years.   Almost all of their observatories are established and operated by local governments for the public service, in what they call a public observatory.   The number of public observatories in Japan is more than 250.   Their telescopes are getting bigger and bigger every year. Their detectors and other instruments are also becoming high grade.   In 2001, a three-year construction program for a 2-m telescope, which is largest

T.D. Oswalt (ed.), The Future of Small Telescopes in the New Millennium, Vol. I, 147–165.

telescope in the domestic facilities, started at the Nishi-Harima Astronomical Observatory, one of the public observatories. The activities of public observatories are attracting much public attention to the astronomy community. The close relation and cooperation among all of the observatories has become important.

1.1   Present Status of the Small Telescopes in Japan

Optical telescopes have contributed to the progress in astronomy and they will continue to do so in future.

In Japan, research astronomy is carried out mainly in the NAOJ, the Institute of Space and Astronomical Science (ISAS), and universities. The NAOJ is the most active center, especially for the observational optical astronomy. Observational optical astronomy has made rapid progress since the establishment of the OAO with 1.88-m reflector in 1960.

Teaching astronomy is carried out mainly in science museums, planetariums and public observatories. Public observatories have been rapidly increasing from the mid-1980s. It is desirable to get out of the education depending on the planetarium and to aim at the real experience. At present, there are 20 telescopes (public observatories are half of them) which are larger than 1-m in aperture, and 81 telescopes (public observatories are almost of them) which are larger than 0.60-m (see Figure 1).

The following is a representative list of small optical and/or near-infrared telescopes in Japan[1]:

- *MAGNUM Observatory*, University of Tokyo (National); Haleakala, Maui, Hawaii, USA
  2.00-m optical/NIR telescope. Foci: RC/9,19.
  Fully automated telescope.
  Monitoring observations for active galactic nuclei.

- *Okayama Astrophysical Observatory of NAOJ* (National); Kamogata, Okayama.
  1.88-m optical/NIR telescope. Foci: N/4.9, Cs/18, Cd/29.
  0.91-m optical telescope. Focus: Cs/13.
  0.65-m solar telescope. Focus: Cd/50.
  Multi-purpose NIR camera; OASIS.
  High-dispersion echelle spectrograph; HIDES.
  Optical polarimetry and spectroscopy system: OOPS.
  Photoelectric magnetograph.

[1] Key to telescope types: Cs= Cassegrain, Cd= Coudé, Ne= Newtonian, Ns= Nasmyth, RC= Ritchey-Chretién. Numbers following these abbreviations indicate the *f*/ratio.

- *Advanced Technology Center of NAOJ* (National); Mitaka, Tokyo.
  1.50-m NIR testing telescope. Foci: Cs/12.2, Ns/12.2.
  Development of and experiments with observational equipment. System support for Subaru telescope.

- *Gunma Astronomical Observatory* (Prefectural); Takayama, Gunma.
  1.50-m optical/NIR telescope. Foci: Cs/12, Ns/12.
  0.65-m optical telescope. Focus: Cs/12
  NIR camera, high-dispersion spectrograph, optical CCD camera and photoelectric photometer.
  Public observatory.

- *Communications Research Laboratory* (National); Koganei, Tokyo.
  1.50-m optical telescope. Focus: Cs/18.
  Tracking telescope for satellites.

Figure 1. Distribution of Telescopes with 60-cm or more in aperture.

- *Nagoya University, Optical and Infrared Astronomy Observatory (National)* ; Sutherland, South Africa.
  1.40-m NIR telescope. Focus: $f/9.9$.
  Simultaneous-color Infrared Imager for Unbiased Survey (SIRIUS).
  Site: South Africa Astronomical Observatory (SAAO).

- *Institute of Space and Astronomical Science* (National); Sagamihara, Kanagawa.
  1.30-m NIR telescope. Foci: Cs/18, Ns/18.

- *Rikubetsu Astronomical Observatory* (Town); Rikubetsu, Hokkaido.
  1.15-m optical telescope. Foci: Cs/8, Ns/8.
  Spectrograph, photoelectric photometer and CCD camera.
  Public observatory.

- *Nakagawa Observatory* (Town); Nakagawa, Tokushima.
  1.13-m optical telescope. Focus: Cs/10.
  CCD camera.
  Public observatory.

- *Misato Observatory* (Town); Misato, Wakayama.
  1.05-m optical telescope. Focus: Cs/8.
  Spectrograph and CCD camera.
  Public observatory.

- *Kiso Observatory*, Institute of Astronomy, University of Tokyo (National); Mitake, Nagano.
  1.05-m Schmidt telescope. Focus: $f/3.1$.
  CCD camera for survey observations.

- *Saji Astronomical Observatory* (Village); Saji, Tottori.
  1.03-m optical telescope. Foci: N/4.3, Cs/6.7, Cd/10.2.
  Photoelectric photometer and CCD camera.
  Public observatory.

- *Agematsu Observatory*, University of Kyoto (National); Agematsu, Nagano.
  1.02-m NIR telescope. Focus: Cs/15.

- *Bisei Astronomical Observatory* (Town); Bisei, Okayama.
  1.01-m optical telescope. Foci: Cs/12, FCs/12.
  Spectrograph, photoelectric photometer and CCD camera.
  Public observatory.

- *Kawabe Astronomical Observatory* (Town); Kawabe, Wakayama.
  1.00-m optical telescope. Focus: Cs/10.
  CCD camera.
  Public observatory.

- *Bisei Spaceguard Center* (Foundation); Bisei, Okayama.
  1.00-m optical telescope. Focus: f/3.
  Automatic observation for near earth objects.

- *Toyama Astronomical Observatory* (City); Toyama, Toyama.
  1.00-m optical telescope. Focus: Ns/8.
  Spectrograph and CCD camera.
  Public observatory.

- *Kagoshima University* (National); Iriki, Kagoshima.
  1.00-m optical/NIR telescope. Foci: Cs/12, Ns/12.
  NIR camera (future plan).

- *Ayabe Astronomical Observatory* (City); Ayabe, Kyoto.
  0.95-m optical telescope. Foci: N/5, Cs/13.5.
  Spectrograph, photoelectric photometer and CCD camera.
  Public observatory.

- *Himeji Hoshinoko-yakata* (City); Himeji, Hyogo.
  0.90-m optical telescope. Foci: N/5, Cs/16.
  CCD camera.
  Public observatory.

- *Nishiwaki Earth Science Museum* (City); Nishiwaki, Hyogo.
  0.81-m optical telescope. Foci: N/5, Cs/15.
  CCD camera.
  Public observatory.

- *Nichihara Observatory* (Town); Nichihara, Shimane.
  0.75-m optical telescope. Foci: N/3, Ns/12.
  CCD camera.
  Public observatory.

- *Sundai Observatory*, Kita-Karuizawa (Private); Naganohara, Gunma.
  0.75-m optical telescope, Foci: N/3, Ns/12.
  Spectrograph, photoelectric photometer and CCD camera.
  School observatory.

- *Nagoya City Science Museum* (City); Nagoya, Aichi.
  0.65-m optical telescope. Foci: N/5, Cs/13.
  Photoelectric photometer and CCD camera.
  Public observatory.

- *Nishi-Harima Astronomical Observatory* (Prefectural); Sayo, Hyogo.
  0.60-m optical/NIR telescope. Focus: Cs/12.
  Spectrograph, Photoelectric photometer, CCD camera and NIR camera.
  Public observatory.

- *Kuma Kogen Astronomical Observatory* (Town); Kuma, Ehime.

0.60-m optical telescope. Focus: Cs/8
CCD camera.
Public observatory.

1.2   National Astronomical Observatory of Japan

The NAOJ is the Inter-University Research Institute for astronomical research in Japan. As the center of astronomical research in Japan, the NAOJ promotes a wide-range of observational and theoretical research and contributes to our understanding and knowledge of the universe.

The NAOJ conducts observations of various astronomical objects and phenomena, including the earth and other planets within the solar system, the sun, stars, galaxies, and beyond to galactic clusters and the fringes of expanding universe.   In addition, the NAOJ contributes to the development and installation of advanced observatory instruments, which will allow the exploration of new phenomena in the universe.

The NAOJ promotes joint-research and joint-use of observatory facilities and instruments by foreign and domestic researchers, in addition to putting efforts into the cultivation of young researchers and graduate-level students. The NAOJ also promotes the wide-range development of astronomy and related fields.   As a facilitator of international co-operation in the field of astronomy, the NAOJ promotes international exchange programs and international joint-research projects. The NAOJ is organized into six research divisions and ten observation/research facilities.

- *Research Divisions*
     Optical and IR astronomy and Observational System Division
     Solar Physics Division
     Astrometry and Celestial Mechanic Division
     Theoretical Astrophysics Division
     Radio Astronomy Division
     Earth Rotation Division

- *Observation/Research Facilities*
     Subaru Telescope
     Mizusawa Astrogeodynamics Center
     Norikura Solar Corona Observatory
     Okayama Astrophysical Observatory
     Nobeyama Solar Radio Observatory
     Solar Activity World Data Center
     Nobeyama Cosmic Radio Observatory
     Astronomical Data Analysis Center
     Advanced Technology Center
     Public Relation Center

The observations by small optical and near infrared telescopes of the NAOJ are taking place at the Okayama Astrophysical Observatory (OAO). The OAO is located on top of Mt. Chikurinji, at southwestern part of Okayama Prefecture, about 200-km west of Osaka in Japan, at an altitude of 372-m. It is the largest optical/IR observatory in the domestic facilities, with three telescopes which apertures are 1.88-m, 0.91-m and 0.65-m.

The sky over the OAO is free from clouds during almost 150 nights per year. There is very little precipitation, the atmospheric transparency is very good, and as the turbulence is generally small, very sharp images may be obtained with the telescopes there.

The 1.88-m reflecting telescope (Figures 2 and 3) was established in 1960. There are three foci, which are: the Newtonian focus with the Imaging Camera, Cassegrain focus with the Spectro-Nebulargraph or the Okayama Astrophysical System for Infrared Imaging & Spectroscopy (OASIS), and Coudé focus with the High Dispersion Echelle Spectrograph (HIDES). The main research areas by the 1.88-m telescope are comets, chemical composition of stars, carbon stars, Be stars, peculiar stars, binary stars, star forming regions, planetary nebulae, novae & supernovae, extra-solar planets, galaxies, active galaxies and clusters of galaxies etc.

Figure 2. The dome of 1.88-m telescope at the Okayama Astrophysical Observatory.
–photo courtesy National Astronomical Observatory of Japan

The 0.9-m reflecting telescope was established in 1960. Its focus is Cassegrain with the Z Spectrograph (prism spectrograph) or the Okayama Optical Polarimetry and Spectroscopy system (OOPS). The main research areas by 0.9-m telescope are asteroids, comets, variable stars, binary stars, BL Lac objects, T Tau objects, young stellar objects, novae & supernova, interstellar dust and spectral classification of stars, etc.

The 0.65-m Coudé solar telescope was established in 1968. It is provided a high dispersion spectrograph, an echelle spectrograph and a photoelectric magnetograph.

The OAO have a future plan which place under the authority of the University of Kyoto as a branch of the Center for Research and Education of Cosmic Activities. The Center's major project is the construction of 3-m new technology optical/IR telescope. The Center will be a base for observations and experiments of related universities, and will make progress for astronomical research and education by the cooperation of related universities. The main research areas of 3-m telescope will be activities of galaxies and cataclysmic phenomena of the universe.

Figure 3. The 1.88-m telescope of the Okayama Astrophysical Observatory.
–photo courtesy National Astronomical Observatory of Japan

2.   UNIVERSITY OBSERVATORIES

The number of university observatories in Japan that conduct astronomical research has decreased. However, some universities have their own research groups and are actively producing high-quality results.

2.1   University of Tokyo

There are two observatories for research at the University of Tokyo.

The Kiso Observatory was established in 1974 as a branch observatory of the Tokyo Astronomical Observatory, University of Tokyo. On 1988, the Kiso Observatory was reformed as a university observatory belonging to the Center for Astronomical Education and Research, Faculty of Science, University of Tokyo. Researches done at the Kiso Observatory are mainly carried out with the 1.05-m Schmidt telescope. Facilities are open to astronomers in both domestic and foreign institutions. The method of observation since the opening of the Kiso Observatory was photographic imaging only. However recently a CCD camera with 2040×2040 pixels has been available. The camera covers a wide view of 50'×50' on the focal plane of the Schmidt telescope. The PtSi near-infrared camera has been available since 1996.

The MAGNUM Observatory with 2-m telescope was established in 2000 at the Haleakala High Altitude Observatory Site on the Island of Maui, one of the Hawaiian Islands as facilities of the Research Center for the Early Universe, School of Science, the University of Tokyo. The MAGNUM Observatory is operated fully automatic. The Multicolor Active Galactic Nuclei Monitoring (MAGNUM) project carries out long-term monitoring of quasars and supernovae in optical and near-infrared regions, and luminosity distances are derived and compared with those predicted from the red shifts of the objects.

2.2   Nagoya University

South Africa 1.4-m telescope was established in 2000 at the South Africa Astronomical Observatory, it is supported by the Optical and Infrared Astronomy Laboratory, Division of Particle and Astrophysical Sciences, Graduate School of Science of Nagoya University. This project carries out near-infrared survey observation in southern-hemisphere region of the sky. Research by the 1.4-m telescope is mainly carried out using the Simultaneous-3 colors Infrared Imager for Unbiased Survey (SIRIUS), which is developed by members of the Laboratory.

2.3   Kyoto University

The Hida Observatory was established in 1968 at Mt. Ooamami, Gifu Prefecture, as a university observatory belonging to Graduate School of Science, Kyoto University. There are three telescopes: a 0.60-m refractor, a 0.60-m reflector and a 0.60-m domeless solar telescope. The fields of observational research conducted at the Hida Observatory are solar system and solar physics.

2.4   Kagoshima University

The 1-m optical and near-infrared telescope of the Kagoshima University was established in 2000 at Iriki town, Kagoshima Prefecture. The main object of this telescope is the observation for luminosity and period of Mira type variables. It will soon house a new distance scale to co-operate with VLBI Exploration of Radio Astronomy (VERA).

3.   PUBLIC OBSERVATORIES

Until 1980, the number of the public observatories in Japan was a few, and most of the telescopes were apertures of under 0.50-m. In the past twenty years, the number of the public observatories is rapidly increasing and the trend for larger telescopes is continuing. Furthermore, the detectors and other instruments are becoming high-grade. A small number of these telescopes are equipped with cooled CCD cameras, spectrographs and photoelectric photometers, with which some academic research is being performed. It is important for cooperation among public observatories and research institutes. The contents of cooperation are information networks, co-observations and personnel exchanges.

3.1   Activities of Public Observatories

Modern activities of the public observatories in Japan have been initiated in 1990, mainly by the establishment of the Nishi-Harima Astronomical Observatory (NHAO). In the circumstances that the activities of the public observatories were mainly star watching, the NHAO (Figure 4) was opened for the public and devoted to public education and astronomical research. It is equipped with two main telescopes and some small telescopes. One of these telescopes is a 0.60-m reflector (Figure 5) equipped with a cooled CCD camera, a spectrograph, a photoelectric photometer and a near infrared camera. Using the 0.60-m telescope, a star watching party is held for visitors

Figure 4. The Nishi-Harima Astronomical Observatory.

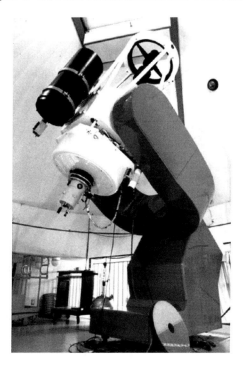

Figure 5. The 0.60-m telescope of the Nishi-Harima Astronomical Observatory.

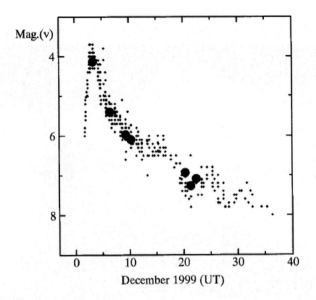

Figure 6. Light Curve of Nova V1496 observed by Narusawa et al.

every night. After 9 p.m. it is used for research and educational programs. At present, the CCD camera is mainly used for imaging of planets, comets, star clusters and galaxies, for photometry of asteroids, variable stars, supernovae and galaxies (see Figures 6 & 7).

The spectrograph is mainly used for spectral classification of stars. The photoelectric photometer is mainly used for photometry of eclipsing binaries. The near infrared camera (Figure 8) is mainly used for imaging of H-II regions and galaxies (see Figures 9, 10 & 11).

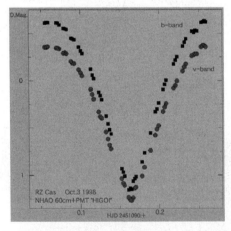

Figure 7. Primary minimum of RZ Cas observed by Narusawa et al.

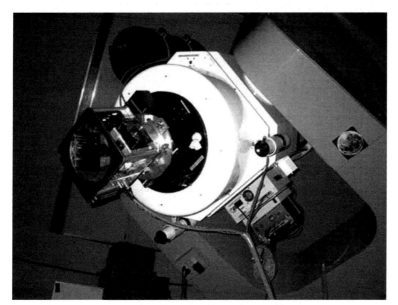

Figure 8. Near Infrared Camera (Nishi-Harima Cosmos: NIHCOS).

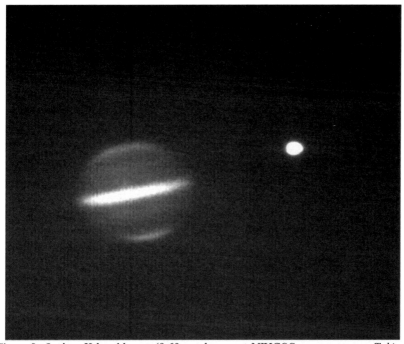

Figure 9. Jupiter: K-band image (0.60-m telescope + NIHCOS.   *–courtesy Tokimasa*

Figure 10. Saturn: K-band image (0.60-m telescope + NIHCOS).   *–courtesy Tokimasa.*

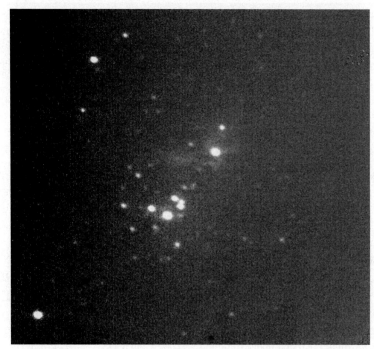

Figure 11. KL Orionis: K-band image (0.60-m telescope + NIHCOS.   *–courtesy Tokimasa*

The other telescope is 0.50-m solar telescope. It is a heliostat type telescope without an observing dome. A monochromatic image is taken through a Lyot filter, passing only H-α light, and a spectrum image is taken through a solar spectrograph. Their images are recorded to videotape, and they are presented through the TV displays in an exhibit hall.

After three years from the establishment of the NHAO, in 1993, the Bisei Astronomical Observatory (BAO) was established as a local center of cultural activities in Okayama area. The main telescope of the BAO is a 1.01-m reflector, which was the first telescope larger than 1-m aperture here. It is equipped with a cooled CCD camera, a grating spectrograph and a photoelectric photometer. The BAO is active in spectroscopy for comets, early type stars and supernovae, especially.

Construction of 1-m class telescopes in Japan continued every year during 1993-1999. In 1999, the Gunma Astronomical Observatory (GAO) was designed for both astronomical research and public use, was established by Gunma Prefecture. The main telescope of the GAO is a 1.50-m reflector, which is largest telescope in the public observatories for the present. It is equipped with a near infrared camera (HgCdTe array of 1024 x 1024 pixels), a high dispersion spectrograph (echelle grating) and a CCD camera. Another telescope of the GAO is a 0.65-m reflector. It is equipped with optical CCD cameras, photoelectric photometer and a middle-low dispersion spectrograph.

Based upon fundamental philosophy of providing every visitor with a real experience, the GAO is also engaged in educational activities that are linked to school or life-long education to spread astronomical observation, as well as observational research activities at the GAO.

3.2   A 2-m Telescope Project of the Nishi-Harima Astronomical Observatory

The atmosphere at the NHAO is transparent, and a beautiful starry sky is frequent, because it is far from large cities. The brightness of the night sky at the zenith is 21.5 mag/arcsec^2 under favorable conditions. This area has small amount of precipitation, and there are about 120 clear nights a year. The scintillation of stellar images due to air turbulence at the NHAO is extremely small, and seeing will sometimes be as sharp as 0.5 arcsec.

Because of the excellent atmospheric conditions at the NHAO, in 2001 a three-year construction program for a 2-m telescope system was started. The system will be used for research in the following areas, using the power of the largest telescope in Japan:

- *Morphologies of nebulae star clusters and galaxies.*
- *Colors of stars.*

Figure 12.   Counterglow in the western part of Scorpius, showing the darkness of the
night sky at the NHAO   *–photo by Ishiguro et al.*

- *Spectra of stars.*
- *Imaging solar system objects* such as planets, comets and asteroids.
- *Imaging distant extra-galactic objects* such as groups of the galaxies.

This system will be designed for real academic research in astronomy and
science education when combined with forefront observational instruments.
Furthermore, the system should be used for cooperative work with other
educational and academic institutes, and with amateur astronomers.

To achieve these goals, the optical and mechanical design includes:

- *The effective aperture of the primary mirror is 2-m.*
- *Ritchey-Chrétien optical design.*
- *Cassegrain focus and two Nasmyth foci.*
- *Near infrared camera and optical CCD camera* at Cassegrain focus.
- *High-sensitivity color video camera and naked eye observation system*
  at one Nasmyth focus.
- *Optical spectrograph* at the other Nasmyth focus.
- *An integrated control system*, which is operated without replacing each
  instrument by switching focus, in order to operate with minimum
  personnel for public, academic research and educational use.
- *Altitude-azimuth mounting.*

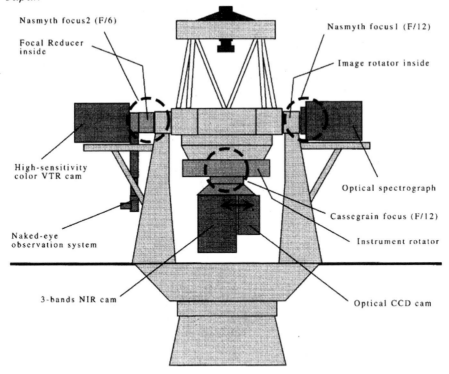

Figure 13. Conceptual design of the NHAO 2-m telescope.   *–courtesy Tsumuraya*

There are five instruments for various observations.

- *Naked-eye observation system* (sliding up and down eyepiece): Instrument for visual observation is devised so that visitors, including physically handicapped people, watch image of object without difficulty. Furthermore, this focus is open to the amateur astronomers and general visitors for attaching of a small photo-camera or a CCD camera. Watching real universe through such and powerful mirror with 2m aperture is very impressive comparing with reading the picture book or watching indirectly through the television.

- *High-sensitivity color video camera* (3 image intensifiers + 3 CCDs): This camera has the same performance with the camera on the SUBARU telescope, which is used to get images for public release. Utilizing the power of the largest telescope in Japan, this camera should be used for making the news release images when suddenly interesting astronomical phenomena occurred. Furthermore, using the color movie in high sensitivity and high quality, it is expected to release original educational materials in multimedia format.

- *Three-band simultaneous near infrared imaging camera* (see Figure 14; J,H,K bands + HgCdTe arrays of 1Kx1K pixels + polarizer + grizm): It will be the first camera in the world, which can perform speckle imaging, polarimetry and spectroscopy in 3-bands, respectively. It is expected to obtain the clear infrared pseudo-color image by subtracting the effect of the fluctuation in the air. This instrument will provide observations of the birth of the stars and the formation of the planetary systems, a panoramic view of our Galaxy and galaxies, and clues to the chemical evolution of the universe.

- *Optical CCD camera* (2Kx2K pixels): It will be used for optical imaging and wide-field imaging. It is expected that not only the amateur astronomers and researchers but also the general people will request to use this camera. This camera will offer the opportunity to obtain the high quality and high-resolution astronomical image of the planets, nebulae, star clusters and comets, which is not accessible by a widely used usual CCD camera.

- *Optical spectrograph* (Czerny-Turner type/ middle-low dispersions): It is used for spectroscopic observation in the optical region. It is expected that real-time changes in spectra can be observed with this instrument.

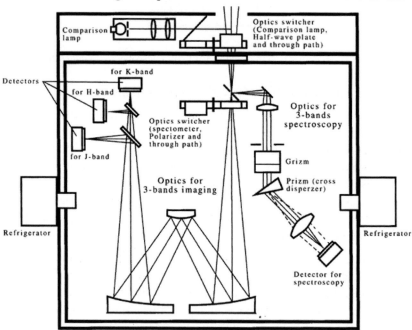

Figure 14. Design concept of the NHAO 3-band simultaneous near-IR imaging camera.
–courtesy Tsmuraya

Computers operate the 2-m telescope system. The operating system on the computer will have easy-to-operate hardware/software interfaces and consistency with other observational instruments and the telescope enclosure. This system, a so-called the integrated control system, has many functions which preserve the control commands, file the data on celestial objects, drive the control of telescope, operating instruments and storage of observational data.

In 2004, the 2-m telescope system will be operational. The research activities by the 2-m telescope will be essential to promote live science education opportunities. This real experience in doing science is a unique part of Japan's public use and support of telescopes.

Chapter 15

Simultaneous and Parallel Use of Small Telescopes for Astronomical Photometry

Kenji Tanabe
Okayama University of Science
Okayama JAPAN

1.   INTRODUCTION

Astronomy has largely been depending on the biggest class telescopes because of their light (photon) collecting power.   However, recent technological advances in optical systems, detectors, and computers have changed the situation: it is possible to construct compact yet sophisticated and powerful personal observatories equipped with those highly advanced instruments.   The main reason is of course CCD detectors' high quantum efficiency.   As a result, even the smallest telescopes can play important roles — a 10-cm telescope equipped with CCD camera can play the role of a 1-m telescope with a traditional photographic camera.   However those that are bigger than 0.5-m are not always suitable for personal use because of the difficulty of self-maintenance.   Here I try to elucidate a possibility of small (<0.5-m) telescope combinations, which may differ from the traditional use.

The idea of simultaneous use (to observe different objects at the same time) and parallel use (to do the same object by a different telescope equipped with different kind of detectors) has stemmed from my experience in student education at the undergraduate level.   In my early experience in student education, in the late 1980s, our detector was a photo-multiplier tube (PMT) photometer attached to a 25-cm telescope.   Its limiting magnitude was about 10.   To observe fainter, more interesting objects like cataclysmic variables the only way was to request time on the 36-inch telescope at the Okayama National Astrophysical Observatory (OAO), 50-km away from my university.   This is the only telescope for photometry in Japan.   Although the Okayama prefecture (located in the southwestern part of Japan) is said to be the best site for astronomical observations, visiting observers frequently suffered from the changeable weather.   In the mid-1990s, astronomical CCD

T.D. Oswalt (ed.), The Future of Small Telescopes in the New Millennium, Vol. I, 167–173.

cameras for stellar photometry became available. In 1997, at IAU General Assembly in Kyoto, Professor B. Warner suggested to me the use of CCD for high-speed photometry instead of PMT. Together with several monographs, the article written by Professor J. Patterson on "the Center for Backyard Astrophysics" in *Sky and Telescope* had an influence on me. By 1998, I had constructed two observatories, one in my own backyard and another on the flat roof of our university building. These two observatories have played a complementary role for variable star observation. At present these two now serve collaborating works with other groups in Japan. The next plan is to have other observatories on the roof of the building and my backyard. The major purpose of having multiple telescopes is for students to perform observations on the same fine night for their own objects. Also, I will expand this scheme to obtain much more intensive and effective observations of the variable stars.

2.   PRESENT SITUATION IN JAPAN

2.1   Available Telescopes, Mounts and Domes

Japanese manufacturers have made an effort to produce high-quality astronomical telescopes for both amateur and professional use. In these ten years Japanese productions have become as excellent as Western ones. It is remarkable that most of these Japanese manufacturers are small enterprises comparable to popular camera makers such as Nikon, Canon and so on. Some of them produce telescopes of high enough quality for professional use. Several makers also produce excellent equatorial mounts. However, very few items seem to be equipped with good enough pointing software for professional use.

We also have several astronomical dome makers. Most of them are small companies also. The problem is that their productions are quite old and conservative. For this reason it is necessary for us to give them special orders or ask for additional changes occasionally. However if you choose a remote-controlled observatory, any type will do.

2.2   Detectors

We have no suitable CCD cameras for astronomical photometry. Japanese productions are excellent but rather for amateur use and taking color photographs. On the other hand, PMT photometers are available for professional use.

2.3   Computers

By the mid-1990s, the most popular PC had been the PC98 series of the NEC production because of its ability to deal with Japanese (Chinese) characters. These were not compatible with IBM PCs. Nowadays Windows machines have already replaced the NEC PC98. Accordingly, software for controlling PMT photometers and telescope pointing based on this machine need to be revised for Windows or PC-UNIX.

2.4   Environments - Climate and Light Pollution

There exist several special geographical and social points. For example:
a)   Changeable weather (not suitable for photometry)
b)   High humidity, especially in summer
c)   Strong wind in northeast Japan surrounding Tokyo.
d)   Light pollution: There exists no site 100-km away from a town

Accordingly, it is almost impossible to find out a good site for astronomical observations of faint objects like nebulae and galaxies. However, for stellar photometry, Okayama is said to be in the best site. In the age of the PMT photometer, it was often said that there were no sites suitable for stellar photometry in Japan. However, the use of CCDs has changed the situation. At any rate, the inhabitants of this district like me have an advantage and responsibility for astronomical observation in Japan (and in Asia).

2.5   Examples

The specifications of my two observatories are as follows:

a)   Observatory on the roof of the university seven-floor building (Figure 1)

Figure 1. Dome on the top of the University building (left) and control room

Inside the dome, a 21-cm Cassegrain (Dall-Kirkham) telescope and equatorial (German) mount are set up. Both of them are Takahashi productions. These are remote-controlled from the room on the highest (7th) floor of the building, with more than ten PCs (Linux and IRAF are installed) for data reduction.

b)   My "backyard" observatory of the author (Figure 2)

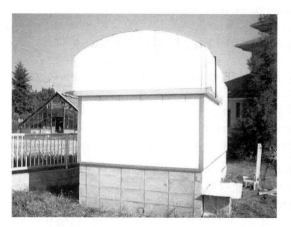 

Figure 2. My backyard observatory and 0.3-m telescope for CCD Photometry

The telescope (30-cm, Dall-Kirkham) and mount are the productions of Takahashi. The present CCD used with this telescope is the ST9E model of SBIG (USA). This observatory consists of a roll-off roof-type observation room and a computer control room. The site is sufficiently distant from central Okayama.

3.   PRESENT PROJECTS

3.1   Education Programs for Students

I usually have both undergraduate and graduate students who choose to study astronomy and astrophysics. They do not always want to be professional astronomers. However, astronomical observations, data acquisition and processing are very good (instructive) and exciting studies. I usually choose for the students one or more objects that are cataclysmic variables and other types of variable stars. I prefer to examine the established results for undergraduate students. In addition, the parallel observation of the well-known variables by CCD and PMT photometer is important for determining the reliability of the CCD data.

3.2   Collaborations with Other Groups

In the past, my observations focused on cataclysmic variables and related objects, independently from other groups. Recently however, collaborations with the Kyoto University team of variable star observers started. The first and second objects were 1RXSJ232953.9+062814 (Uemura et al. 2002) and DM Draconis (Kato et al. 2002), whose superoutbursts were detected in November 2001. Moreover, recent space-based observations discover new cataclysmic variable stars that need to be observed optically (getting the light curves and detecting the orbital periods). Another possible collaboration is the observation of X-ray binaries (black hole candidates) together with space-based observations.

3.3   Other Projects in the Near Future

At one time I had several projects that were to be accomplished using a PMT photometer at OAO. One of them was to detect the period change of AM Canum Venaticorum stars and other short orbital period binaries. However the PMT photometer attached to 36-inch telescope had been removed for renewal. Also, I had started another project to detect the orbital periods of old novae. Unfortunately the interface of the TI-CCD attached on the OAO 36-inch telescope was broken due to a lightening strike. At present, it is almost hopeless to repair this system because of financial limitations. However, these projects will be moved to our personal observatories.

4.   TECHNOLOGICAL PROBLEMS

As I mentioned above, small (less than 0.5-m) telescopes and mounts are usually easier to maintain. However, various kinds of problems arise when we set up an observing system by assembling off-the-shelf items. For example:
a) Pointing software often is not for professional use. It is desirable to install more sophisticated software with a telescope analysis program.
b) Pointing precision is not good for long focal length (large f-number) telescopes due to gear deficiencies. Installation of a friction drive system is one of the best ways to improve this.
c) Generally speaking, telescopes and mounts on the market (ready-made) are not very robust, except the products of very few manufacturers. In particular the attachment to the CCD is usually very weak.

d) In the summer, the temperature becomes too high for some electronic cooling systems to handle.

To overcome these problems, it is necessary to have collaboration between users and manufacturers. Even so, owning multiple telescopes and detectors is occasionally useful, as a back up is then readily available when one malfunctions.

5.   PROSPECTS IN THE NEXT 10 YEARS

5.1   Use of Optical Fibers

Telescopes smaller than 0.5-m often cannot support big detectors, such as a nitrogen-chilled CCD photometer. Use of optical fiber is considered to be one of the best ways to solve this problem. This is thought to be a modern version of the Coudé focus. If it is introduced, we will able to do not only high precision (low-noise) photometry but also spectroscopy and polarimetry.

5.2   Robotic (Automated) Observation

It is often said that the recent weather in Japan is more changeable than it was about 20 years ago. Under such circumstances, the introduction of the robotic telescope is one of the best ways for ground-based observation.

We have not yet tried to introduce an automated photometry system because we lack a reliable weather sensor. I believe that the construction of a sophisticated remote observing system is more important for our observations.

5.3   Remote Stations Inside and Outside of Japan

It costs some million yen (about ten thousand dollars) to build a small telescope and observatory. Hence, to construct one at a long distance is not too expensive. Southern Hemisphere sites can provide much more fruitful observing. If no resident operator is needed, the Internet may provide access to the telescope. My idea is to have an observatory at some place in Australia (if possible) in order to facilitate observations from June to August (these are the worst months here in Japan).

6.   SINGLE USER, MULTIPLE TELESCOPES - EPILOGUE

It is plausible that from the recent space-based observations and other projects an increasing number of new variable stars (mainly cataclysmic variables) and other strange stars will be detected.   Moreover, true photometric nights will become more and more precious.   So in order to perform photometric observations here in Japan it is necessary for us to set up multiple telescopes for a single observer.   This may be a drastic but practical change of the style of the astronomical observation.   I think this will bring great progress in understanding the nature of the variable stars.

7.   ACKNOWLEDGEMENTS

I wish to express gratitude to Dr. Terry Oswalt for giving me the opportunity to introduce my observing system and projects.

8.   REFERENCES

Henden, A.A. & Kaitchuk, R.H. 1990, *Astronomical Photometry*, William-Bell
Howell, S.B. 2000, *Handbook of CCD Astronomy*, Cambridge
Kato, T., et al. 2002, IBVS No. 5284.
Martinez P., & Klotz, Alan 1998. *A Practical Guide to CCD Astronomy* Cambridge 1998
Patterson, J. 1998 *Our Cataclysmic Variable Network*, Sky and Telescope, October, 77-81
Uemura, M. et al. 2002, Publ. Astron. Soc. Japan 54, L15-18

Chapter 16

International Aspects Between Central Europe and Central Asia

Juraj Zverko
Astronomical Institute, Slovak Academy of Sciences
Tatranská Lomnica SLOVAKIA

Abstract: A short historical view of evolution of astronomy and a current status in the countries included under influence of the former Soviet Union is given. A few topics ideal for 'small telescopes' in solar and stellar physics and Solar system astronomy are discussed. A list of observatories in the countries between Central Europe and Central Asia and their instrumentation is given.

Key words: former Soviet Union, international collaboration, small telescope priorities

1.   INTRODUCTION

Astronomy, in its essence, is a more international science than any other one. As to small countries, moreover, international cooperation has been of vital importance for recent several decades. Undoubtedly this will be true in the future. I will concentrate in the following on my experience in the former "socialist block", especially the former Soviet Union states and Central Europe where I have been doing my science since the late sixties of the past century.

2.   ASTRONOMY IN THE PAST AND INTERNATIONAL COLLABORATION

Beginnings of astronomy (together with other natural sciences) in these countries reach back to the 15th century. Substantial development came in the first half of the last century followed by progress benefiting from the start of the '*cosmic age*'. Observatories were established and equipped with

T.D. Oswalt (ed.), The Future of Small Telescopes in the New Millennium, Vol. I, 175–187.
© 2003 *Kluwer Academic Publishers. Printed in the Netherlands.*

telescopes of mainly German and Russian provenience. Exclusive of the 6-m *'big glass'* their size ranges from 2.6- to about 0.6-m, though a number of smaller telescopes have been exploited for scientific, educational and public aims. While the 0.6-m reflector from Carl Zeiss Jena (CZJ) might be the most common type of telescope in these countries, the biggest one, 2.6-m and a few automatic *'1-m class'* telescopes were designed and produced by the Leningrad Optical and Mechanical Factory (LOMO).

The birth of modern astronomy in Slovakia dates back to the third decade of the 20th century when a 60-cm Newton/Cassegrain reflector was installed in *Stará Ďala.* In 1943 a new high-altitude (1780-m elev.) observatory, the *Skalnaté Pleso,* in mountainous part of Slovakia, was established by Antonín Bečvář and equipped with the telescope which had been moved here from Stará Ďala. In 1962 a coronal station was opened on top of the 2633-m elev. high *Lomnický štít* and equipped with a 20-cm coronagraph. In 1987 a solar horizontal spectrograph and a 0.6-m reflector were put into operation near *Tatranská Lomnica* at the foot of the hill. Observational programs for the Sun, stars and interplanetary matter were being run.

Science was relatively well supported by governments in Czechoslovakia in third quarter of past century and a new 2-m telescope was built in 1967 at Ondřejov near Praha. The need for various, mainly spectroscopic, observations encouraged making connections abroad. So far, many papers have been published in collaboration with foreign astronomers and which exploit the observational data obtained by them. The collaborations often became institutionalized by means of bilateral agreements between national scientific centers, but this was not always the rule. These bilateral agreements serve as a tool for collaboration with countries in the 'West'. The collaboration between the 'East' countries was organized under "The Multilateral Collaboration of Academies of Sciences of the Socialist Countries" while universities collaborated within their own programs. Though this agreement could only be as effective as single institutions were able to make it, it represented a significant contribution, as it enabled a huge exchange of astronomers and extended observational possibilities.

3.   CURRENT SITUATION

Everywhere in the world observatories have been built up, usually at urban and university centers, where preceding astroclimatic observations have either not been performed or were not considered of high importance. As the urban centers grew and the accuracy of astronomical observations increased, the observatory sites were shown to be unsuitable and newer ones were built in higher altitudes on mountains, further from the centers. This elevating of telescopes brought improved limpidity; however, the seeing strongly depends on local conditions. For example, while the Skalnaté Pleso Observatory has very poor seeing in summer time, excellent conditions are reported year-round from Central Asian mountain observatories. While sophisticated focal plane instruments are able to cope with poor seeing and rapidly variable local atmospheric extinction, the general lower limpidity over the old lowland observatories cannot be avoided. This is so for example in the Ondl ejov Observatory near Praha, the Czech Republic or the Crimean Astrophysical Observatory, Ukraine. For the Central European climate moreover, variable weather and higher humidity is characteristic. The instability is limiting for the quality of photometric observations performed with a classic single-channel photoelectric photometer. This is illustrated in Zverko (1982): differences between two consecutive photometric measurements range from 0.003 mag in good conditions to 0.02 mag in poor ones. Conditions, however, may change during a night. Nevertheless, there are still tasks to be successfully accomplished using these telescopes and their instrumentation. Observations became more effective and valuable when obtained in coordinated campaigns (Kreidl et al. 1990), Weiss et al. (1998) or long-term observational projects, (see Hric et al. 1996).

4. HIGH PRIORITY PROGRAMS

High priority opportunities for small telescopes are summarized comprehensively in many other chapters of this book. Besides the afore-mentioned coordinated and long-term campaigns, I would like to mention some additional problems that are especially suitable for small telescopes.

The *'20-cm class'* Solar telescopes with coronagraphs should, among other projects, aim at (i) monitoring and forecasting active solar events; and (ii) the mechanism of coronal heating, the nature of solar wind and coronal mass ejections' acceleration (see also Rušin & Rybanský 1982).

As to stellar astronomy/astrophysics I would emphasize studies of the structure of stellar atmospheres for which "stellar telescopes" with advanced instrumentation are able to give high precision photometric, polarimetric and spectroscopic data for stars: (i) the complicated nature of chemically peculiar stars, their atmospheric abundances and structures, the radiative and dynamic processes in magnetic field; (ii) the structure of atmospheres of tidally distorted components of binary stars, gravitational limb darkening and circumstellar structures in binary/multiple stars. A sophisticated software package, FOTEL (Hadrava 1990), has been prepared for such studies. Also a code for the solution of light and radial velocity curves from high precision photometry and spectroscopy is available. KOREL (Hadrava 1995) is a code for disentangling multiple spectra and Doppler imaging.

Finally, (iii) asteroseismology is a tool to study inner structure of stars and a test of evolutionary theories.

Solar system astronomers will certainly find these telescopes effective in the Near Earth Objects (NEO) and projects as (i) modeling of shapes of asteroids based on photometry in relation to their origin and evolution, (ii) physics of comets and their activity in large distances from the Sun.

5.   THE LIST OF TELESCOPES

Observatories and their telescopes are summarized below. Even though we list here more than one hundred telescopes, the list remains incomplete. While the upper limit of "small telescopes" is easily recognized at 2.6-m, the lower one is spread below 0.5-m, as valuable photometry has been done using photoelectric photometers and/or CCDs.

Many observatories in the countries concerned maintain incomplete websites or even stay silent when addressed by e-mail. Nevertheless, the list can be a good guide for those who would intend to take advantage of excellent observing conditions mainly in the Central Asia region and start collaboration. Some of the observatories have already been participating in the Whole Earth Telescope network (Mt. Suhora, Poland, Mt. Maidanak, Uzbekistan). Four observatories created "The Central Asian Network": Odessa, Ukraine, Mt. Dushak-Erekdag, Turkmenistan, Mt. Tien-Shan, Kazakhstan and Abastumani, Georgia.

Regretably, in the end of recent century this region was branded by wars in former Federal Republic of Yugoslavia. Cultural and scientific establishments were destroyed. For example, the Astronomical Observatory of the Sarajevo University on Mt. Trebevič, Bosnia and Hercegovina, was completely destroyed, including precious photographic plate archives, library, even the 0.65-m Newton/Cassegrain and 0.4-m Cassegrain telescopes.

In the following sections, notice the wide instrumental basis of astronomy in Poland that also runs a 1.3-m telescope on Las Campanas, Chile. Also Ukraine represents a strong astronomical community. In their observatories at Odessa and Crimea new telescopes and instrumentation have been built. The four Central Asian countries offer excellent observational conditions and operate on bi- or multilateral basis with institutions mainly in former Soviet Union. Some of the observatories need material support as well as encouraging in their effort to maintain their institutions. This mainly concerns the institutions in former Yugoslavia and some in former Soviet Union states. But all the staffs are professionally prepared to accept collaboration.

5.1 The Czech Republic

5.1.1 Astronomical Institute, The Academy of Sciences of the Czech Republic, Observatory Ondʔejov, 528-m elev.

2-m Coudé/Cassegrain telescope (CZJ), focus: grating spectrograph with Reticon (alternatively CCD) 3 cameras enabling reciprocal dispersion from 4Å/mm to 16Å/mm in *UB* region; Cassegrain focus: HEROS two-channel echelle spectrograph, 3700-5750Å, CCD 2000 x 800 pixels, 22 μm/pixel, 5800-8350Å, CCD 1152x770 pixels, 15 μm/pixel; usual observing conditions: from 1 arcsec to poor seeing; stellar spectroscopy

0.5-m horizontal Solar telescope with high-dispersion spectrograph; Solar physics

0.2-m solar telescope, H-α filter, video recording and picture analysis, multicamera spectrograph; evolution of active regions and eruptions

5.1.2 Astronomical Institute, Charles University, Prague, Observatory Ondʔejov

<u>0.65-m Cassegrain</u> reflector, CCD; stellar photometry, NEO asteroids

5.1.3 Kle> Observatory, 1007-m elev.

1-m reflector
0.57-m reflector
0.63-m Maksutov; research field: new asteroids, NEO, comets

5.1.4 Astronomical Institute, Masaryk University, Brno

0.6-m Newtonian, CCD; stellar photometry

5.2 Poland

5.2.1 Astronomical Observatory of the Jegellonian University, Fort Skala near Krakow

0.5-m Cassegrain (CZJ)
0.35-m Maksutov

5.2.2 Warsaw University Astronomical Observatory, Ostrovik Observatory, 200-m elev.

0.6-m Cassegrain (CZJ), CCD, limiting magnitude 18-19
(1.3-m Cassegrain, Las Campanas Observatory, Chile)

5.2.3 Torun Center for Astronomy, Nicolas Copernicus University, Piwnice,

0.9-m Schmidt-Cassegrain
0.6-m Cassegrain (CZJ)

5.2.4 Poznan Astronomical Observatory, Adam Miczkiewicz University

0.35-m Cassegrain

5.2.5 Wroclaw Astronomical Observatory, Wroclaw University, Bialkow

0.6-m Cassegrain (CZJ), CCD
horizontal Solar telescope
coronagraph

5.2.6 Mt. Suhora Astronomical Observatory, Pedagogical University, Krakow, 1009-m elev.

0.6-m Cassegrain (CZJ), dual-channel photometer, CCD

5.3 The Slovak Republic

5.3.1 Astronomical Institute, Slovak Academy of Sciences

- **Skalnaté Pleso Observatory, 1780-m elev.**
0.60-m Cassegrain (CZJ), *UBVR, uvby* photoelectric photometer; stellar photometry
0.61-m Cassegrain, CCD; asteroids and comets
- **Stará Lesná Observatory, 800-m elev.**
0.60-m Cassegrain (CZJ), *UBV* photoelectric photometer; stellar photometry
0.50-m horizontal Solar telescope with high-dispersion fibre optic spectrograph, TV data picture recording; Solar photosphere
- **Lomnický Štít Coronal Station, 2633 elev.**
0.2-m double coronagraph, CCD camera and high-speed TV photometer; solar chromosphere and corona;

5.3.2 Astronomical Institute, Comenius University, Observatory Modra-Piesky, 500-m elev.

0.6-m Newtonian/Cassegrain reflector (CZJ), CCD; asteroids

5.3.3 Public observatories in Slovakia

In the frame of significant support to cultural activities by the socialist government in the past era many public observatories were established, some of them with good telescopes:

- County Observatory Humenné, Kolonica station, Eastern Slovakia
1-m Cassegrain, spectrograph (donated by the Odessa Observatory)

- Observatory Hlohovec, Western Slovakia
0.6-m Cassegrain (CZJ), CCD; stellar photometry

5.4 Hungary

5.4.1 Konkoly Observatory, Hungarian Academy of Sciences, Piskestetö, 958-m elev.

1-m RCC (CZJ), CCD
60/90-cm Schmidt (CZJ)
0.5-m Cassegrain

5.4.2 Gothard Astrophysical Observatory, Ötvös Lóránd University, Szombathely, 216-m elev.

0.6-m Cassegrain (CZJ), CCD

5.4.3 Heliophysical Observatory, Hungarian Academy of Sciences, Debrecen, 132-m elev.

0.53-m coronagraph, H-α filter

5.4.4 Baja Observatory

0.5-m RC

5.4.5 Szeged University Observatory

0.4-m Cassegrain, CCD; stellar photometry and astrometry

5.5 Croatia

5.5.1 Hvar Astrophysical Observatory, University of Zagreb

0.65-m Cassegrain reflector for stellar electrophotometry *UBV*; stellar photometry
1-m Cassegrain

5.6 Slovenia

5.6.1 Astronomical Observatory, University of Ljubljana, Crni Vrh

0.36-, 0.25-, 0.19-m Schmidt-Cassegrain, CCD; stellar photometry, limiting magnitude 18 to 16 magnitude, respectively.

5.7 Yugoslavia

5.7.1 Astronomical Observatory in Belgrade

0.16-m astrograph, CCD; minor planets, comets

5.8 Bulgaria

5.8.1 National Astronomical Observatory Rozhen, Bulgarian Academy of Sciences, 1759-m elev.

2-m RCC (CZJ), Coudé spectrograph, CCD;
0.6-m Cassegrain (CZJ); photoelectric photometry
0.5/0.7-m Schmidt

5.8.2 Astronomical Observatory Belogradchik, Bulgarian Academy of Sciences, 610-m elev.

0.6-m Cassegrain, CCD, stellar photometry

5.9 Romania

5.9.1 Astronomical Institute, Romanian Academy of Sciences

- Bucharest Observatory
0.5-m Cassegrain (CZJ), photoelectric *UBV* and CCD photometry
0.38-m double astrograph, CCD
0.13-m solar refractor, *H-α* filter

- Cluj-Napoca Observatory
0.51-m Newtonian, UBV photoelectric photometry
0.4-m Schmidt/Cassegrain

- Timisoara Observatory
0.3-m Cassegrain, photoelectric photometry

5.10 Estonia

5.10.1 Tartu Observatory

1.5-m Coudé/Cassegrain (automatic AZT-12, LOMO) Cassegrain and
Coudé spectrographs, CCD; stellar physics
0.6-m Cassegrain, CCD

5.11 Lithuania

5.11.1 Moletai Observatory, Institute of Theoretical Physics and Astrophysics, 220-m elev.

1.65-m reflector, three-channel photoelectric photometer, CCD
0.63-m reflector
0.35/0.51-m Maksutov

5.12 Ukraine

5.12.1 Main Astronomical Observatory, National Academy of Sciences of Ukraine, 180-m elev.

0.7-m reflector, photometry, polarimetry, spectroscopy of Solar system
bodies and stars
0.4-m astrograph, BV photographic, limiting magnitude 15th; star's
proper motions, Galactic structure, positions of Solar system bodies
0.4-m double wide-angle astrograph (CZJ), limiting magnitude 16th;
planets, asteroids, comets
0.45-m double horizontal Solar telescope

5.12.2 Ukrainian-Russian Centre of Astronomical and Medical-Ecological Investigations, Peak Terskol, Caucasus, Russia, 3100-m elev.

2-m RCC telescope (CZJ), echelle Coudé spectrograph, CCD, IR
Fourier spectrometer; galactic and extragalactic objects
0.8-m telescope

Two 0.6-m Cassegrain (CZJ) telescopes; photometry and spectroscopy
of stars and comets
0.7-m horizontal solar telescope, CCD, Solar spectra

5.12.3 Astronomical Observatory, Odessa State University

0.8-m RC reflector
0.6-m Cassegrain
0.5-m Cassegrain, photometer *UBVR*
0.48-m Cassegrain, CCD, *UBVIJK*
0.35-m Schmidt
    Photoelectric and CCD photometry, spectroscopy; periodic
    and aperiodic processes in stars, galactic chemical evolution,
    asteroseismology, comets

- Mt. Dushak-Erekdag Station, Turkmenistan
0.8-m RC reflector
0.6-m Cassegrain

5.12.4 Astronomical Observatory, Kharkov National University, Grakovo, 156-m elev.

0.7-m AZT-8 reflector; solar photosphere, asteroids, comets, NEO
0.5-m reflector
0.20-, 0.27-m refractors

5.12.5 Mykolayiv Astronomical Observatory, 52-m elev.

0.15-m meridian circle
0.12-m zone astrograph

5.12.6 Crimean Astrophysical Observatory, Nauchny, 600-m elev.

2.6-m Cassegrain/Nasmyth/Coudé (LOMO), Coudé spectrograph 2.4-12
    Å/mm, 4000-9000Å, CCD 1024x280 pixels 20μm/pixel, for K0,
    V=9 mag, *R*=30000, *S/N* >100 at 20 min; Nasmyth spectrograph 35
    Å/mm, 3500-10000Å, CCD 600x480 pixels, 20 μm/pixel, for 14
    mag (Seyfert galaxy), *R*=7000, *S/N*>100 at 60 min; variable stars,
    stellar atmospheres, galaxies
*1.25-m RC (*LOMO), five-channel *UBVRI* photometer/polarimeter, 0.001
    sec time resolution
1-m RCC telescope (CZJ), Simeiz station
0.5-m meniscus, BVR stellar photometer, slitless spectrograph, a point
    source of 14 mag

0.4-m double astrograph (CZJ), 6.9° objective prism/direct imaging; small bodies of Solar system

Ground-based gamma-ray telescope GT 48, 52 m², observations of very high energy (> 10^{11} eV) sources

1.2-m tower solar telescope, Cassegrain spectrograph with double magnetograph and vector magnetograph; high-degree helioseismological network Mt. Wilson-Crimea-Kazakhstan

0.6-m tower Solar telescope, universal spectrophotometer, CCD

0.53-m Lyot coronagraph, spectrograph

0.21-m coronagraph, *H*-α filter, magnetic field of the Sun and sunspots, imaging of the Sun in He I, solar flares

5.13 Russia

5.13.1 Institute of Astronomy, Russian Academy of Sciences

- Simeiz station, Crimea
1-m RCC telescope (CZJ

- Zvenigorod station, 173-m elev.
1-m RCC telescope (CZJ)

5.13.2 Special Astrophysical Observatory, Russian Academy of Sciences, Nizhnnij Arkhiz 2070-m elev.

1-m RCC telescope (CZJ), echelle spectrograph; *UBVRI* CCD photometer, limiting magnitude 20, supporting telescope of the 6-m

0.6-m Cassegrain telescope (CZJ); supporting telescope for the 6-m

5.13.3 Pulkovo Observatory, Russian Academy of Sciences, 75-m elev.

- Pulkovo Observatory station, Russian Academy of Sciences, Kislovodsk
0.5-m coronagraph

5.13.4 Sternberg Astronomical Institute, Moscow University

- Crimea station
1.25-m telescope (LOMO)
0.6-m Cassegrain (CZJ)
0.5-m Cassegrain (CZJ)

5.13.5 Engelgardt Astronomical Observatory, Kazan State University, 98-m elev.

0.48-m Maksutov
0.48-m Cassegrain AZT-14 (LOMO), photoelectric photometer
0.4-m astrograph

- South Astronomical Station, Antalia, Turkey, 2485-m elev.
1.5-m RCC AZT-22, imaging CCD system

5.14 Armenia

5.14.1   Byurakan Astrophysical Observatory, 1500-m elev.

1/1.3-m Schmidt
2.6-m Cassegrain/Nasmyth/Coudé; physics of stars and nebulae, extragalactic objects

5.15 Azerbaijan

5.15.1   Shemakha Astrophysical Observatory, 1400-m elev.

2-m RCC (CZJ); Coudé echelle spectrograph, CCD, cooperation with the Nicolaus Copernicus University, Torun, Poland

5.16 Georgia

5.16.1   Abastumani Astrophysical Observatory, Georgian Academy of Sciences, 1700-m elev.

1.25-m RC (LOMO), two-star photometer uvbyβ
0.7-m Maksutov, CCD,
0.48-m reflector, photometer UBVR
0.4-m double astrograph (CZJ)
0.4-m refractor
Solar telescopes

5.17 Kazakhstan

5.17.1   Tien-Shan Observatory, 2800-m elev.

2 x 1-m RCC (CZJ), spectrograph, Fabry-Perot interferometer, two-channel polarimeter, photometer *WBVR*
0.8-m reflector (CZJ)
2 x 0.48-m Cassegrain, photometer *WBVR*

0.5-m horizontal Solar telescope (CZJ), spectrograph, large 'Nikolsky' coronagraph

5.17.2 Alma-Ata Observatory

0.6-m Cassegrain (CZJ), CCD

5.18 Uzbekistan

5.18.1 Ulugh-Beg Astronomical Observatory, Uzbek Academy of Sciences, Mt. Maidanak, 2600-m elev.

1.5-m RCC AZT-22, (LOMO)
1-m RCC (CZJ)
2x 0.6-m Cassegrain (CZJ)
2x 0.48-m Cassegrain

5.19 Turkmenistan

5.19.1 Mt. Dushak-Erekdag Observatory, 2020-m elev.

1-m wide-field telescope, CCD
0.8-m RCC, two-star photometer *UBVRI*
2x 0.5-m Cassegrain

5.20 Tadjikistan

5.20.1 Sanglock Observatory, Institute of Astrophysics, 2300-m elev.

1-m RCC (CZJ)
0.7-m RC (LOMO)
0.6-m Cassegrain (CZJ)

6.   REFERENCES

Hadrava, P., 1990, Contrib. Astron. Obs. Skalnaté Pleso 20, 23
Hadrava, P., 1995, A&AS 114, 393
Hric, L., Skopal, A., Urban, Z. et al. 1996, Contrib. Astron. Obs. Skalnaté Pleso 26, 121
Kreidl T. J., Garrido, R., Huang, L. et al. 1990, MNRAS 245, 642
Rušin V., Rybanský, M., 1982, Bull. Astron. Inst. Czechosl. 33, 219
Weiss, W. W., Kuschnig, R., Mkrtichian, D. E. et al. 1998, A&A 338, 919
Zverko, J., 1982, Bull. Astron. Inst. Czechosl. 33, 314

Chapter 17

GNAT—A Global Network of Small Astronomical Telescopes

David L. Crawford, Eric R. Craine and Roy A. Tucker
GNAT, Inc., Tucson, Arizona USA and
Goodricke-Pigott Observatory, Tucson Arizona USA

Abstract: New generation small telescopes are powerful tools for astronomical research and for educational outreach. GNAT, the Global Network of Astronomical Telescopes, is a non-profit organization whose goals are to implement and operate a global network of relatively small telescopes. GNAT should have nearly infinite upside potential for astronomers and students everywhere. Its Value Per Cost Ratio is enormous.

Key words: telescopes: automated, networks

1.   INTRODUCTION

Small telescopes have always been powerful tools for astronomy, and the new technology aspects that have become available in recent years will only amplify this value. New generation small telescopes, ones using modern CCD imaging detectors and taking full advantage of computer control and communication networking, can be truly frontier instruments for astronomical research.

These new generation aspects apply as much for small telescopes as for the largest telescopes. However, the relative cost of a large telescope to that of a small one is very large. (See the Appendix on Astro-economics as food for thought.)

It is a fact that the automatic and networked use of such small telescopes could provide many more quality observing hours to the astronomical community worldwide, both for research and for education, and it would allow programs to be done which are quite difficult or impossible now.

T.D. Oswalt (ed.), The Future of Small Telescopes in the New Millennium, Vol. I, 189–200.
© 2003 *Kluwer Academic Publishers. Printed in the Netherlands.*

A non-profit organization, GNAT, Inc., has been created with the goals of developing and operating a global network of such small telescopes and of being a catalyst for those interested in the effective use of small telescopes for research and for education.

2.   GNAT

GNAT has developed specifications for such small new generation telescopes, particularly for one of 0.8-m aperture. These specifications are given in the Appendices. In addition, GNAT is involved in a collaborative effort to develop and operate small "scan-mode" telescopes with the goal of using them in such a global network.

GNAT's goal is to be a relatively low cost operation, with low overhead and a small staff, but with many members, allies, and partners. We believe that this program has essentially zero risk coupled with unlimited upside potential for research and for educational outreach. Funding it is the only issue remaining—a problem in common with most projects.

Current planning involves:

1. *Obtain a lease/option on such a 0.8-m telescope.* This size of telescope is expected to be the GNAT standard. This telescope will be used to prove system efficiency and to investigate all aspects of routine operation. Lease/option arrangements are ideal for such developmental telescopes, allowing full testing of the telescope mechanical and control systems, without obligation to exercise the option unless full performance specifications are obtained. When these specifications are met, the option will be exercised and the telescope purchased.

2. *Continued development and then construction of a Scan Mode Facility.* This project involves implementation of a network of 14-inch aperture telescopes designed by Roy Tucker, which are operated in an electronic scan mode. This very modest cost system has several virtues in addition to its low cost:

   a) faint limiting magnitude possible (the prototype, now in operation is doing near Earth asteroid work to 20th magnitude);

   b) easy replication to multiple sites;

   c) easy operation and maintenance;

   d) low housing costs;

   e) very high data throughput in a network; and

    f) Ideally suited to a variety of synoptic, long term monitoring projects (for example, extra solar planet searches).

3. *Development of the GNAT Web site* to be a viable resource both for GNAT itself and for most small telescope users.

4. *Host meetings on topics of interest relative to GNAT.* The most recent of these resulted in the publication of a volume in the A.S.P. Conference Series (Craine, Tucker & Barnes 1999).

5. *Funding development and operation.* As with all telescope projects, especially small telescopes, obtaining funding is a key problem. However, the "Value-to-Cost" ratio is enormous. Many astronomers and students will benefit. Such a facility is a key component of a balanced funding program for astronomy, and its use will greatly complement larger telescopes and other astronomy facilities.

In some sense, the viable mix of telescope sizes in astronomy is like that of operating a truck fleet. One must have trucks of all sizes, so as to efficiently match the size of the truck to the need. It is the only cost effective way to do business, or science.

3. GNAT PROJECTS NOW UNDERWAY

There have been several engineering and science programs undertaken by GNAT during its early years. These programs were put in place to accomplish four goals: 1) to evaluate the SciTech 0.5-m automated telescope, 2) to gain further practical experience in CCD photometry in order to better plan GNAT science programs, 3) to undertake improvements in GNAT telescope control software as well as data reduction algorithms, and 4) to enable publication of results geared toward supporting efforts to expand interest in the GNAT program.

Several of these programs have been structured to address multiple goals. The primary observing programs to date include open cluster monitoring, with particular emphasis on M67, Mira and semi-regular variable monitoring, and potential extra-solar planetary transit searching. Each of these programs has taught us a great deal about limitations imposed on the projects by the SciTech telescope, and has led to a better understanding of the current state of issues surrounding commercially available automated telescopes. Each has also helped immeasurably in establishing data reduction and analysis protocols for future GNAT programs.

In addition, we have been fortunate to be able to involve several students in GNAT projects which has not only contributed to the publication of several

GNAT papers, but has also allowed us to begin to develop a protocol for incorporating students in GNAT programs. These efforts have included students from high school, undergraduate and graduate programs. Publications of student efforts include Craine (1996), Culver & Craine (1997), Giampapa, Craine & Hott (1995), Craine, Crawford & Craine (1998), Craine (1999), Craine, Crawford & Craine (1999), Taylor, Craine & Giampapa (1999), Roberts, Craine & Giampapa (2000) and Barentine (2001).

One of the findings of this body of work has been that commercial automated telescopes suitable for GNAT purposes are still in a fragile state of development. The concept of "laying your money down, taking a system out of the vendor catalog, and starting to collect data" appears to remain a remote one. This has led us to consider an additional option for GNAT data collection: the small, automated drift scan systems of a type under development by GNAT member Roy Tucker, and for which the prototype development is described here.

The past several years have seen remarkable advancement in the participation of amateurs in the field of astronomical research, primarily due to the easy availability of good CCD cameras and ever more powerful computer equipment. As search programs extend to ever-fainter objects, both amateurs and professionals must refine their methods to improve both magnitude penetration and area coverage. A prototype instrument consisting of three identical scan-mode telescope/camera combinations has been constructed and will help explore how this improvement in capability may be economically accomplished.

The actual instrument telescopes are conventional Newtonians with 35-cm aperture, *f*/5 primaries of low-expansion Astrositall material and 8-cm minor axis secondaries of fused quartz. A temperature-compensating optical support structure using the differential expansion of steel and aluminum rods eliminates the need for focus changes during the course of an evening's observations. Since the telescopes are intended to be directed towards a fixed azimuth and elevation for months at a time, there is no problem with structural flexure due to changing orientation and a relatively lightweight structure has been implemented.

Another unusual aspect of the telescope design is that focusing is accomplished by moving the primary mirror. This is a somewhat tedious process but fortunately, due to the temperature compensation mechanism, is seldom needed. First, the primary mirror is adjusted so as to be perpendicular to the optical axis. Three dial gauges indicate the positions of the three adjustment screws on the primary mirror. During the focusing process, the differences in the readings are maintained so as not to change the collimation while identical increments of displacement are applied to the three screws.

Approximately an hour is required to fully accomplish the iterative focusing process.

To facilitate directing the instrument at the desired point in the sky, each telescope is supported by a short yoke English mounting fabricated from commonly available structural steel and plumbing fittings. There is no clock drive mechanism or precision slow motions other than some threaded steel rods and nuts to rigidly hold the instrument in the declination and hour angle positions. At those rare times when the instrument pointing is changed, those steel nuts may be "tweaked" with a wrench to effect a certain amount of coarse adjustment. One full rotation of the declination adjustment nut effects a motion of 5.85 arc-minutes. Fine positioning is accomplished using dial gauges to measure motions of the telescope primary mirror cell relative to the observatory floor: 0.001-in corresponds to six seconds of arc.

Three home-brew cameras based upon an existing design (Tucker 1995) have been constructed using thinned, back-illuminated 1024x1024 CCD imagers from Scientific Imaging Technologies (Beaverton, Oregon). These TK1024 devices have 24-micron pixels and, in combination with the telescope system, produce an image scale of 2.83-arcseconds per pixel. The resulting field of view is 2898-arcseconds or 48.3-arcminutes or 0.805-degrees.

These cameras are capable of being operated in continuous scan mode and the equivalent integration time at the celestial equator is approximately 193-seconds. The combination of integration time and aperture reveals stars as faint as 20.5 magnitude. Scanning at the celestial equator permits examination of just over 12 square degrees per hour. In normal operation, the three telescopes are aimed at the same declination but spread in right ascension at intervals of 15 to 60 minutes to produce a data stream of image triplets separated in time that reveal moving and time-varying objects. At this time, the instruments are centered on +03 degrees, 18 minutes, and 20 seconds declination. The separation between the three instruments is about 20 minutes of right ascension.

The presence of unattenuated bright moonlight greatly reduces the dynamic range of the images, perhaps even producing saturation of the CCD. Before each camera is a filter slide mechanism that permits insertion of a clear filter of anti-reflection-coated BK7 glass or a color filter of photometric "V" or "I". Consideration was initially given to the use of neutral density filters to attenuate the light levels during times of bright moon, but it was quickly realized that color filters produce the desired attenuation and also provide useful color photometry. The filters are arranged so that image triplets ordered in time in "I", "V", and "I" are obtained, permitting interpolation of an "I" magnitude at the time that the "V" image is acquired.

The instrument shelter is a conventional roll-off roof design, eight by twelve feet in size, optimized for the enclosure of three instruments that are pointed in the general direction of the celestial equator. Eight-foot high walls that provide protection from artificial light sources in the vicinity of the facility restrict the view of the sky. An electrically powered garage door opener is used to automate the motion of the roof. Although this mechanism extends across the center of the opened aperture, the telescopes can easily be positioned to avoid any obscuration of the optics. Fundamental considerations in the design of the shelter were low cost and unattended operation.

This type of instrument has no expensive precision mounting. The telescopes are fixed in position and scan the same strip of sky each night. The cost of the instrument is primarily in the optics and the cameras, those components that are required for the detection process. A precisely pointing and tracking mount, while very expensive, does not detect a single photon.

Tucker envisions multiple such instruments surveying parallel swaths in the equatorial region of the sky night after night for use in near Earth asteroid detection programs. For GNAT purposes, higher time resolution is obtained by simply adding more telescopes into the network—a feasible approach given the very low cost of the telescopes. This type of instrumentation can benefit from the savings that can be realized from assembly line mass production. It is also very forgiving of component failure since a problem that would render a single telescope of a five-telescope instrument inoperative would simply mean that one was then using a four-telescope instrument with a somewhat higher threshold of object detection.

4.   IS THERE A NEED FOR GNAT?

Yes. Major observatories worldwide have been closing down their small telescopes, even very productive ones, so as to concentrate on their large telescope programs. University astronomers and students have heavily used most of those small telescopes. What can they do now? Some have their own facilities, but most do not. Some have access to a facility at a good observing site, but most do not, nor ever will. All do need access to a good telescope, with good instrumentation, at a good observing site—both for themselves and for their students.

There are many scientific programs that such telescopes can do. In fact, they are better for these programs than are larger telescopes. These include monitoring programs of all sorts, including any objects variable on most any time scale (variable stars, galaxies, and other objects), open cluster photometry, standard systems and calibrations, and on and on.

5.   WHY GNAT SPECIFICALLY?

We have been dialoging with astronomers about their needs for some years now, visitors at the major observatories, at scientific meetings, and elsewhere. We are very interested in the technology and the applications. We like and have experience with a wide variety of small telescopes. Very few astronomers at universities have the time or the experience to develop either their own facility or a networked one. They barely have time to do the actual research (and the proposals).

Here is our scenario for the next decade and beyond: As noted above, astronomers and students at most universities (USA and otherwise) do not have access to a professional observing facility. Those that do have little time to spend in telescope development and operation, and they also have the continual problems of adequate operating funds (as do the large facilities at major observatories), of access, and of liability and modernization.

GNAT is a way to solve these problems. It can be used by all, producing a lot of data for all astronomers and for their universities, without any local operational problems. Clearly this is a situation where one plus one is much more than two. By working together, we can each accomplish much more, and at much less cost and bother.

Think about the amount of quality data that a minimum of twelve telescopes located at at least six worldwide quality observing sites could produce, with well more than 200 good observing nights per year at each site, a good CCD imaging photometer on each telescope, automatic observing mode, and with standardized filters and observing procedures.

A number of these telescopes would undoubtedly operate also with other instrumentation, such as spectrometers. It is also easy to add capability to the network, with other telescopes, sites, and instrumentation.

There is a real need for a GNAT. With GNAT, astronomers do research and GNAT takes care of the rest: development, implementation, operation, liabilities, and all the non-research aspects of owning and operating an observatory. All members would, however, be active partners in the effort. When funded, it would be a most powerful, distributed, facility for use by many astronomers from many institutions in many countries.

It appears to us to be the most cost-effective way to implement small telescopes at good sites for the entire worldwide community over the next decade.

6.   A FEW QUOTES FROM THE GNAT BYLAWS CONCERNING GNAT'S OBJECTIVES:

1. Service to the public and to astronomy via scientific research, education, and public information concerning the development and operation of a global network of (small) astronomical telescopes for astronomical research and education, for developing national and international cooperation, and for archiving and standardizing essential astronomical data and information.
2. Service to the membership and others via collection of information, distribution of information, education on all aspects of the use of small telescopes and related topics and assistance with member's problems by sharing knowledge on a local, national, and international basis.
3. Any moneys or funds from membership, programs, or services shall be used totally for research and education, and to further the goals of GNAT. No funds shall be used for personal gain except for reimbursement of expenses for program activities.
4. GNAT has been organized exclusively for educational and scientific purposes within the meaning of Section 501(c)(3) of the United States Internal Revenue Code. Its FIN is 86-0712639.

7.   SUMMARY

There are many aspects to the issues of small telescopes as part of a balanced and rational approach to astronomical research and education. We have just outlined a few of them here. We are sure that the technology and the need and the potentials are all ripe for a viable global network of small telescopes. GNAT is one mechanism to help make it happen. There is no doubt in our minds that it can be done or that it should be done. It is of very low risk, of relatively low cost, and of "astronomically" high potential. The Value-to-Cost ratio is enormous.

We hope that all those interested individuals and organizations will join us in helping to bring GNAT to a reality. We welcome and need your active involvement.

8.   REFERENCES

Barentine, J.C. 2001, M.S. Thesis, Colorado State University, Ft. Collins, Colorado.
Craine, P.R. 1996, GNAT Tech. Rep. TR96-1101, Global Network of Astronomical
    Telescopes, Tucson, AZ.

Craine, P.R. 1999, in "CCD Precision Photometry Workshop", ASP Conf. Series Vol. 188, 133-142.

Craine, E.R., Crawford, D.L. & Craine, P.R. 1998, in "Observatory Operations to Optimize Scientific Return", P.J. Quinn (ed.), Proc. of SPIE Vol 3349, 335-342.

Craine, E.R., Crawford, D.L. & Craine, P.R. 1999, in "CCD Precision Photometry Workshop", ASP Conf. Series Vol. 188, 83-94.

Craine, E.R., Tucker, R.A., Barnes, J. (eds.) 1999, "CCD Precision Photometry Workshop," ASP Conference Series No. 189, ISBN 1-58381-015-3

Culver, R.B. & Craine, P.R. 1997, IAPPP Comm. 68, 49-59.

Giampapa, M.S., Craine, E.R. & Hott, D.A. 1995, Icarus 118, 199-210.

Roberts, J.H., Craine, E.R. & Giampapa, M.S. 2000, BAAS 32, 1459.

Taylor, J.M., Craine, E.R. & Giampapa, M.S. 1999, in CCD Precision Photometry Workshop, ASP Conf. Series Vol. 188, 238-251.

Tucker, R.A. 1995, BAAS 185, 63.04.

9.   APPENDIX A: ASTRO-ECONOMICS

How do some interesting things scale with telescope aperture (Aperture to some power)? Do some of these things relate to a telescope's value? Abt, Trimble and others (including us) have looked at a number of these items, and some discussion of such is given elsewhere in the present volume.

1. *Photons collected?* How much light are we collecting for analysis?

2. *Number of users?* How many individuals can a telescope service, or handle? A subset of this question might be "How many students?" Another subset might be "What is the distribution of how many nights were assigned for the research program?" Another might be "Distribution of the number of team members, if more than one?" For example, does a 10-m serve three times that of a 1-m? Ten times?

3. *Number of papers* resulting from the research?

4. *Number of citations* to the papers? Does a paper done with results from a 10-m have three times the citations of one from a 1m? Ten times? Thirty times? On the average, of course, as not all papers are created equal.

5. *Telescope cost?* As built, not as it was proposed.

You decide on what exponent seems to make sense. Make your estimates before you read on.

Table 1 below gives some of the calculations that one might find useful to consider. We have asked a number of people over the years about items 2 to 4 and have found a great range in estimated exponents, from negative to about 1.5. Very few have been over that value. Most have thought that the exponent for item 2 is less than for item 3 and that the one for item 3 is less than for item 4.

As to cost, all previous studies (when "all things are equal") have shown that this value is less than 3 but more than 2.5. You can see in the table how even a small difference can lead to a big difference in costs for a large telescope. It is important to be aware of this and to try to hold such escalations as low as possible (short focal ratio of the primary, for example, as long as the value does not in itself make a big impact on cost). It is clear that this look relative to cost should use at least two terms, one being about 2.8 and the other one of lower value (1 or 1.5?) for those items that do not scale at such a high rate. The term with the lower exponent has little effect for the largest apertures, of course. Astro-economics is an interesting subject, though not much studied. And so are astro-psychology and astro-sociology. All of these disciplines are interesting and important when thinking of the viability of small telescopes.

Table 1.

Spreadsheet showing the result of looking how entries escalate with different values of the exponent. A is the aperture of the telescope in meters.

Aperture A	A^2	$A^{0.5}$	$A^{1.0}$	$A^{1.5}$	$A^{2.5}$	$A^{2.8}$	A^3
0.5	0.3	0.7	0.5	0.4	0.2	0.1	0.1
0.8	0.6	0.9	0.8	0.7	0.6	0.5	0.5
1.0	1.0	1.0	1.0	1.0	1.0	1.0	1.0
1.5	2.3	1.2	1.5	1.8	2.8	3.1	3.4
2.0	4.0	1.4	2.0	2.8	5.7	7.0	8.0
2.5	6.3	1.6	2.5	4.0	9.9	13	16
3.0	9.0	1.7	3.0	5.2	16	22	27
3.5	12	1.9	3.5	6.5	23	33	43
4.0	16	2.0	4.0	8.0	32	49	64
6.0	36	2.4	6.0	15	88	151	216
8.0	64	2.8	8.0	23	181	338	512
10.0	100	3.2	10	32	316	631	1000
15.0	225	3.9	15	58	871	1964	3375

10. APPENDIX B: Standardized Design Parameters for a GNAT 0.8-meter Telescope

1. Imaging and photometry are the main roles for the telescope, but not the only ones.

2. An imaging CCD photometer is the main instrument, but one or two other ports exist.

3. Value per Cost is a critical item. High quality at low cost is the goal.

4. /N and 80/20 and 1+1>2 are important issues.

5. Keep it simple.

6. Uniformity in design and fabrication and software is essential. Standardize.

7. Reliability is critical. Low maintenance costs are as important as low capital costs.

8. Alt-Az and Equatorial are both acceptable designs for NGT small telescopes.

9. The primary mirror focal ratio is to be in the range 1.5 to 2.0.

10. The secondary focal ratio is to be in the range 6 to 9.

11. A possible flavor is to have a considerably wider field telescope (PF?).

12. Field of View (FOV): Design for a potential 4000 sq CCD, implement for a 2000 sq.

13. Match pixel size, seeing, and field of view.

14. Image quality: Between 0.6 and 0.8 and-sec FWHM. Smooth surface.

15. Pointing: Open loop: Approx 10 arc-sec. Closed loop: Approx 1 arc-sec.

16. Tracking: Approx 0.1 arc-sec over several minutes.

17. Telescope control system (TCS) to allow for full automatic and remote operation.

18. Telescope scheduling. Multi-user, multi-telescope software needed.

19. Documentation.

20. Housing and site issues: Mainly a local issue, and not in the standardized design.

11. APPENDIX C: Specific Discussion Issues at the Past GNAT Meetings:

1. Alt-AZ vs. Equatorial? Either is acceptable.

2. The FOV, f-ratio, pixel matching compromises?

3. How many "flavors"?   Cost and performance impacts of varying the specs?

4. Data acquisition, data handling, processing, communications, archiving?

5. Other telescope issues?

6. Other GNAT issues?

Chapter 18

The International Perspective:
From Small Astronomical Telescopes to the World Space Observatory

Hans J. Haubold
Office for Outer Space Affairs, United Nations
Vienna, AUSTRIA

Willem Wamsteker
European Space Agency, IUE Observatory
Villafranca, Madrid SPAIN

Abstract: The UN/ESA Workshops on Basic Space Science are a long-term effort for the development of astronomy and regional and international cooperation in this field on a worldwide basis, particularly in developing nations. The first four workshops in this series (India 1991, Costa Rica and Colombia 1992, Nigeria 1993, and Egypt 1994) addressed the status of astronomy in Asia and the Pacific, Latin America and the Caribbean, Africa, and Western Asia, respectively. One major recommendation that emanated from the first four workshops was that small astronomical facilities should be established in developing nations for research and education programs at the university level and that such facilities should be networked. Subsequently, material for teaching and observational programs for small optical telescopes were developed or recommended and astronomical telescope facilities were inaugurated at UN/ESA Workshops on Basic Space Science in Sri Lanka (1995), Honduras (1997), and Jordan (1999). Elements of the Workshops, focusing on teaching, observing programs, and the Japanese donation program for small astronomical telescopes are briefly summarized in the first part of this paper.

The second part of this paper is a report on the recent UN/ESA Workshop on Basic Space Science, held at the University of Mauritius in 2001, and a summary of achievements of the UN/ESA Workshops on Basic Space Science for the period 1991 to 2001. Since 1991, similar reports, issued for each of the UN/ESA Workshops on Basic Space Science, have been brought to the attention of UN Member States on an annual basis with the objective to gain more support for the worldwide development of astronomy.

Key words: United Nations, UN/ESA Workshops, Basic Space Science, small telescopes

T.D. Oswalt (ed.), The Future of Small Telescopes in the New Millennium, Vol. I, 201–225.

1. ASTROPHYSICS FOR UNIVERSITY PHYSICS COURSES, OBSERVING PROGRAMMES, AND SMALL ASTRONOMICAL TELESCOPES

The locations and dates of recent UN/ESA Workshops on Basic Space Sciences[1] are shown in Figure 1. They focus on the long-term development of astronomy on a regional to international basis, particularly in the developing countries (see Burdyuzha & Khozin 2000; Heck 2001).

1.1 Teaching Module "Astrophysics for University Physics Courses"

A teaching module was developed for the UN/ESA workshops on basic space science. It presents an array of astrophysical problems, any one or a few of which can be selected and used within existing physics courses on elementary mechanics or on heat and radiation, kinetic theory, electrical currents and in some more advanced courses (Wentzel 1998). The module presents an answer to the problem of how to introduce astrophysics in physics courses at the university level, in particular in developing countries. For review and implementation, "Astrophysics for University Courses" was provided to astronomical institutions in Bolivia, Colombia, Costa Rica, Egypt, Ethiopia, Honduras, India, Jordan, Mauritius, Nigeria, Paraguay, the

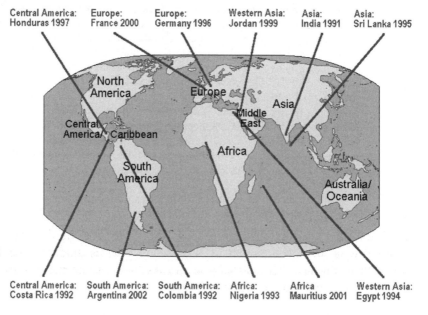

Figure 1. Locations of the UN/ESA Workshops, 1991-2002.

[1] See the UN/ESA Workshops home page at http://www.seas.columbia.edu/~ah297/un-esa/

Philippines, Sri Lanka, Uganda, and Zambia (Wentzel 1998). Also recommended were the materials prepared by Bennett (2001) et al. and Bennett (2002).

Such astrophysics problems are designed to be an interesting and challenging extension of existing physics courses, to determine the student's understanding of physics by testing it in new realms and to stretch the student's imagination. A brief tutorial on astrophysics is provided with each problem so that the physics teacher can present the problem in class. The higher-level problems start with a brief introduction to the physics involved.

Wentzel's teaching module is structured into the following key sections: orbits and Kepler's third law; the solar system; neutron stars and clusters of galaxies; thermal radiation; the lives of stars; cosmic magnetic fields; and high-energy astrophysics. All the problems require compact algebraic and numerical solutions that can easily be translated into physics. For many problems, the solution is shorter than the statement of the problem.

Astrophysics is an attractive science not only because it stretches the imagination but also because it is highly interdisciplinary. Astrophysics involves atomic physics, nuclear physics, fluid and plasma physics, solid state physics, chaos theory, organic chemistry, special and general relativity and more. Students are trained in solving specific problems, however, and they acquire a broad view of science largely through solving many kinds of specific problems. Thus, the problems in this teaching module provide a focus for students to which the broader astrophysical challenges can be linked. Most of the text provided with each problem is designed to highlight the broader questions and challenges, which are then crystallized in the given specific problems that are to be solved by the students.

Even students can formulate good questions suitable for research. Some observations made by the Hubble Space Telescope have been requested and will be investigated by high-school students. However, the frontier nature of astrophysics makes teaching it difficult. Even the professional astrophysicist soon learns to admit to some students' questions that he or she does not know and preferably also to suggest that they study the problem together. Indeed, the problems in the teaching module will be difficult to teach because students will inevitably ask questions that go well beyond the specific problem and the tutorial astrophysics provided. Nevertheless, the value of students' formulating questions far exceeds the discomfort of the teacher admitting that he or she does not know. Many physics students merely memorize their physics. The astrophysics breaks them out of memorization and gets them to think independently. Students' questions are a sign of their progress.

In approaching the theoretical examination of a newly observed phenomenon one should not start with a computer but by determining which kinds of physics are relevant. It is essential to select a few physical parameters and construct a minimum of analytical equations that contain the essential physics. These are often called "back-of-the-envelope" calculations. In astrophysics, one first considers appropriate forms of energy without worrying at the outset about the detailed forces that lead to those energies. One needs to ascertain whether gravitational, nuclear, kinetic or electromagnetic energies or some exchange between two of them are in play and what the main parameters are, such as size or mass of an object, that influence these energies. Sometimes answers can be found by dimensional analysis. It matters little if the numerical coefficients in such estimates may be off by a factor of two or three. Several of the problems in the teaching module emphasize order of magnitude of analysis and dimensional analysis. In particular, some problems ask students to solve differential equations by a one-step integration, which explicitly brings out the main physical parameters.

Frontier science is a collaborative venture. Discussion is an integral part of learning and research in astrophysics. If necessary, the problems in the teaching module can be presented and solved as part of a lecture, but they are selected and written so that they can be discussed and solved by small groups of students, preferably during a class period. Groups of two, three or four students work well, depending in part on the physical limits of the seating arrangement.

Students working in groups take much time. The teacher can lecture about three problems in the time that student groups need to do one. Compared with pure lecture courses, some topics of the course must be omitted because the time is no longer available. Nevertheless, assuredly the students will understand the one problem they solved and the teacher will have evidence of it. This is much more useful to the student, in the long run, than some additional material stored incompletely in his or her memory.

1.2 Observing Programs for Small Optical Telescopes 'Hands-On Astrophysics'

The "Hands-On Astrophysics" (HOU) project uses the variable star database of the American Association of Variable Star Observers (AAVSO). It is a curriculum for high school and college science, math, and computer science classes, and involves students in the scientific process. HOA helps students acquire science skills and develop an understanding of basic astronomy concepts, it provides interdisciplinary connections, and takes students through the whole scientific process while working with real data. The HOA curriculum also informs students about variable stars and their importance to the professional astronomical community, and gives them the necessary information and skills to study variable star behavior. HOA was provided by AAVSO, free of charge, to Ethiopia, Honduras, Jordan, Morocco, Paraguay, the Philippines, Sri Lanka, Uganda, and Zambia.

Students learn the necessary skills to make observations, analyze their data with graphing and statistical techniques, make predictions, and compare predicted and observational values, as well as learn how to develop sophisticated mathematical models. Students learn about variable stars by using the activities, software, charts, slide sets, and videos that accompany the teacher and student manual. Students are able to access the AAVSO database, and share their investigations and observations with other students via the Internet web site specifically developed for this project[2].

The study of variable stars is particularly suited to science, math, and computer education. Students can observe variable stars, and analyze the brightness changes in the stars they observe by using the database of 600,000 observations and the computer programs provided. Except for a few exercises in skill development, there are no "right" answers in this curriculum. The data obtained and the results of the analysis of that data by students are the "right" answer. The amount of data and the mathematical refinement techniques give reasonable accurate results. Students understand that their observations can be reliable, and that their data can be useful enough to be used by professional astronomers.

HOA includes the following material: (1) a teacher/student manual, (2) computer software for PCs (VSTAR: a data plotting and analysis program, HOAENTER: a data entry program, HOAFUN: an introduction to variable stars, including a variable star observing game), (3) 45 variable star charts, (4) 14 prints of the Cygnus star field, (5) 31 slides of 5 constellation areas, and (6) 1 video cassette in 3 segments (backyard astronomy, variable stars, and how to observe variable stars).

1.3 The 45-cm Telescope Facilities Donated by Japan

In 1982, the Government of Japan started to provide scientific and educational equipment to developing nations (Kitamura 1999a/b). This program is called the Japanese Cultural Grant Aid (ODA). The provision of astronomical instruments such as research telescopes and planetaria is part of this program. The application for aid needs to be properly made through the Embassy of Japan in the respective country. Through this program, reflecting telescopes and accessories have been provided to Singapore (1987), Indonesia (1988), and Thailand (1989). Since 1991 the donation of telescopes was facilitated through the UN/ESA Workshops in Sri Lanka (1995), Paraguay (1999), and the Philippines (2000). The accessories include photoelectric photometers, spectrographs (or CCD), and computer

---

[2] http://www.aavso.org/

equipment. The optical part of the telescope consists of a parabolic primary mirror of effective aperture of 45-cm, its focal length is 1800-mm, and thus the system works at $f/4$. A hyperbolic surface of 150-mm effective diameter, with Cassegrain compound focal length of 5400-mm serves as secondary mirror. This mirror is operating at $f/12$. The auxiliary guiding telescope has effective aperture of 80mm with a focal length of 1200-mm.

The software that is provided with the telescope allows (I) pointing of moving celestial bodies such as comets and asteroids, (ii) pointing to more than 10,000 celestial objects, photoelectric measurements and observations of variable stars using a photoelectric photometer, and (iv) automatic trailing during spectroscopic observations.

1.4 Conclusion

The diminishing availability of small astronomical telescopes may pose serious threats to the type of research and educational projects that can be undertaken in the 21st century (see Oswalt 1998; U.S. National Academy Press 2001a/b). Large astronomical telescopes will dramatically increase the number of photons collected per year around the world, but provide much less observing time per astronomer than presently available. If such a trend continues, the world's astronomers will have to content for a relatively small number of increasingly oversubscribed large facilities. The UN/ESA Workshops on Basic Space Science in the last decade of the 20th century have involved astronomers from 124 nations (see Table 2). They have called on governments around the world to provide more support for the establishment and operation of small astronomical telescope facilities for the benefit of astronomy around the world. The UN/ESA Workshops on Basic Space Science, to be organized in the first decade of the 21st century will continue to pursue this objective.

2.   REPORT ON THE TENTH UN/ESA WORKSHOP ON BASIC SPACE SCIENCE

2.1 Introduction

2.1.1 Background and Objectives

The Third United Nations Conference on the Exploration and Peaceful Uses of Outer Space (UNISPACE III) and the Vienna Declaration on Space and Human Development recommended that activities of the United Nations

Program on Space Applications promote collaborative participation among Member States at both the regional and international levels. They emphasized the development of knowledge and skills in developing countries[3].

At its forty-third session, in 2000, the Committee on the Peaceful Uses of Outer Space endorsed the program of workshops, training courses, symposia and conferences planned for 2001. Subsequently, the General Assembly, in its resolution 55/122 of 8 December 2000, endorsed the United Nations Program on Space Applications for 2001[4].

Pursuant to resolution 55/122 and in accordance with the recommendation of UNISPACE III, the Tenth United Nations/European Space Agency (ESA) Workshop on Basic Space Science: Exploring the Universe; Sky Surveys, Space Exploration and Space Technologies was organized by the United Nations, ESA and the Government of Mauritius at the University of Mauritius, in Reduit, Mauritius, from 25 to 29 June 2001. The workshop was co-organized by the Centre National d'Études Spatiales of France, the German Space Agency (DLR), the National Aeronautics and Space Administration (NASA) of the United States of America, the National Astronomical Observatory of Japan and the Planetary Society.

The workshop continued the series of United Nations/ESA workshops on basic space science organized for the benefit of developing countries that was initiated in 1991 (See Table 1).

The main objective of the workshop was to provide a forum to highlight recent scientific results obtained using ground-based and space-borne observatories in studies of the stars and the far reaches of the universe. Satellite missions constitute an impressive means of studying all aspects of basic space science from space as a complement to studies being done from the ground. The question of the large volumes of data generated by such missions was discussed in relation to changing research needs within the scientific community, as was how access to the important databases maintained by major space agencies could be facilitated. The importance of data research and education based on space missions was discussed, together with the relevance of such missions to the needs of developing countries wishing to participate actively in the voyage of discovery through the universe.

The present report was prepared for submission to the Committee on the Peaceful Uses of Outer Space at its forty-fifth session and to its Scientific

---

[3] See *Report of the Third United Nations Conference on the Exploration and Peaceful Use of Outer Space, Vienna, 19-30 July 1999* (United Nations publication, Sales No. E.00.I.3), chap. I, resolution 1, part I, para. 1 (e) (ii), and chap. II, para. 409 (d) (i).

[4] Official Records of the General Assembly, Fifty-fifth Session, Supplement No. 20 (A/55/20), para. 37.

and Technical Subcommittee at its thirty-ninth session. A number of papers presented at the workshop will be published in Seminars of the United Nations Program on Space Applications: Selected Papers from Activities Held in 2001 (ST/SPACE/7).

During the workshop, the National Commission on Space Activities (CONAE) of Argentina announced that it would be prepared to host the Eleventh United Nations/European Space Agency Workshop on Basic Space Science at the Mario Gulich Institute for Higher Space Studies in Cordoba, in cooperation with the University of La Plata, Cordoba, Argentina, from 9 to 13 September 2002.

2.1.2 Program

At the opening of the workshop, introductory statements were made by representatives of the Government of Mauritius, the University of Mauritius, ESA and the United Nations. The workshop was divided into scientific sessions, each focusing on a specific issue. Presentations by invited speakers describing the status of their findings in research and education were followed by brief discussions. Invited speakers presented fifty-two papers from both developing and industrialized countries.

The workshop sessions focused on (1) sky surveys; (2) from solar/planetary systems to galactic/extragalactic systems; (3) data manipulation, databases and multi-wavelength analysis; (4) education with and networking of telescopes, with special reference to the southern hemisphere; and (5) utilization of space science and technologies and their benefits to society. Poster sessions provided an opportunity to focus on specific problems and projects in basic space science.

2.1.3   Attendance

Researchers and educators from developing and industrialized countries in all economic regions were invited by the United Nations and ESA to participate in the workshop. Participants held positions at universities, research institutions, observatories, national space agencies, international organizations and in private industry and were involved in all the aspects of basic space science covered by the workshop. Participants were selected on the basis of their scientific background and their experience with programs and projects in which basic space science played a leading role.

Funds provided by the United Nations, ESA and the University of Mauritius were used to cover travel and living costs of participants from developing countries. Some 65 specialists in basic space science attended the workshop.

The following 28 Member States were represented at the workshop: Austria, Canada, Chile, China, Denmark, Egypt, Ethiopia, France, Germany,

Hungary, India, Italy, Japan, Mauritius, Mexico, Netherlands, Norway, Romania, Russian Federation, South Africa, Spain, Sri Lanka, Syrian Arab Republic, Uganda, United Kingdom of Great Britain and Northern Ireland, United States of America, Yemen and Zambia.

2.2 Observations and Recommendations

The important initiatives arising from previous United Nations/ESA workshops on basic space science and their fruitful promotion in Africa were noted by the participants, as also the importance of the regional centers for space science and technology education affiliated with the United Nations in providing the essential knowledge for promoting various programs in space science and technology.

Participants of the workshop reviewed in four working groups the observations and recommendations of past United Nations/ESA workshops on basic space science in the areas of: (1) space exploration; (2) sky surveys; (3) education, training and services; and (4) space technologies.

2.2.1 Space Exploration

Given the steady progress in space science and technology in the past decades, the scientific goal is the quest for knowledge about the structure and evolution of the universe, in particular, to learn more about the solar system, which is fundamental for humanity. The technical gains, technological challenges and spin-offs of space exploration for both industrialized and developing countries are enormous.

Space exploration by individual developing countries alone is very often difficult to achieve. Cooperation between developing and industrialized countries is essential, in particular in view of the fact that the international space community is witnessing a large influx of data from different space probes where space scientists from developing countries can contribute in an important way.

Associated with the concept of the world space observatory (for the ultraviolet region of the electromagnetic spectrum) (WSO/UV), an international center for astronomy could be established—similar in spirit to the Abdus Salam International Centre for Theoretical Physics in Trieste, Italy—that would provide opportunities to space scientists from developing and industrialized countries alike to undertake joint research projects.

Participants noted that nearly all the data by spacecraft operated by ESA, NASA and the Institute of Space and Astronautical Science of Japan in the field of solar physics were available in public archives. Solarsoft, a solar data analysis package, was widely used. To allow a wider distribution, the participants recommended reviewing the possibilities of developing Solarsoft into a complete package that could be made available eventually as a freeware set of solar data analysis tools.

2.2.2   Sky Surveys

Participants noted the importance of sky surveys.   In particular, the emerging need for multi-wavelength astronomy, ranging from radio, infra-red, optical, X-rays to gamma rays, including neutrino and gravitational wave astronomy, provides a rich opportunity for astronomers from developing countries to be encouraged to carry out research, training and education. South-South collaboration involving the Indian Ocean countries on the rim and African countries could be envisaged; for example, collaboration between the MRT and the Southern African Large Telescope or other optical/radio observatories in southern Africa, could be explored.

Participants welcomed the establishment of an international implementation committee, the World Space Observatory Implementation Committee, for the WSO/UV and the progress made in discussions among several space agencies and interested countries. Further development of the project, including wider participation, should be encouraged.

2.2.3   Education, Training and Services

Participants recognized that the area of space science was of a multidisciplinary nature, since it involved modern technologies in electronics, sensor devices and imaging, information technologies, Internet-based technologies, basic science, analytical techniques and so on. Therefore, all countries should address the introduction of the subject at the primary, secondary and tertiary levels of education in order to face future challenges.   Education,   training   and   research   should   not   be compartmentalized. They were interconnected and neglecting any one of them would lead to deficiency in the development of space science in a country.

Mauritius carried out diverse activities, which could be used as an example for other developing countries on the range of space activities that could be used to develop a rich basic space science program. Such activities were (1) the Mauritius Radio Telescope project (since 1989), being undertaken in collaboration between India and Mauritius (detailed information in section IV below); (2) the Mapping Mauritius Project, using remote sensing and geographic information systems (GIS) (since 1997), in collaboration between the University of Mauritius and Phillips University, Marburg (Germany); (c) the interconnection of education, research and training at the University of Mauritius at both the undergraduate and the postgraduate level; and (d) the development of new tools for handling astronomical data based on image-processing techniques using GIS for the classification of galaxies using dynamic neural networks.

However, one should caution that a particular role model should not be pushed in other countries or regions. What worked in one region at one point of time would not necessarily suit another region at some other point of time. Developing basic space science education should be encouraged. In addition, if adequate research facilities were not available in a country, students would not find required challenges and would have no role models to look up to for further motivation. Thus, careful career planning for scientifically educated people, keeping in mind the local environment, must be an integral part of the development process.

The establishment of an African Institute of Space Science as a distributed organization would act as a source of vision and strategy to promote the development of basic space science throughout Africa and would be a major step towards extending the participation of developing countries in Africa in basic space science, which could possibly in turn accelerate the integration of spin-off benefits of space science into society. The institute could benefit from the previous experience obtained in the regional centers for space science and technology education affiliated with the United Nations. Participants recommended that African Governments, while developing and fostering national basic space science programs, give due consideration to supporting the establishment of an African institute of space science, as appropriate to their needs, and that serious consideration be given to the possible benefits of affiliating national programs to the institute.

The participants noted the great value of NASA's Astrophysics Data System in providing access to astronomical literature. Participants urged industrialized countries to ensure continuing support for free access to the system and urged developing countries to utilize fully the services provided by it.

Having in mind the observations and recommendations of past United Nations/ESA workshops on basic space science, participants discussed the Network of Oriental Robotic Telescopes project and made the following observations and recommendations:

Many countries affiliated to the Network project were progressing in terms of education by developing youth programs, scientific clubs, secondary school programs and courses at university levels;

However, more countries, with well-developed activities in astronomy, astrophysics and space science, should introduce basic space science courses into their university curricula and should train young space scientists, astronomers, astrophysicists, programmers, engineers and technicians in their laboratories or observatories for appropriate periods of time. It was appreciated that the development of cooperative scientific projects through

Ph.D. work and continuous collaborations between universities in developing and industrialized countries was the best way to develop basic space science;

French and Libyan astrophysicists were developing an educational and scientific cooperative program for the 2.3-m national telescope at the University of Benghazi. The Islamic Republic of Iran had joined the Network of Oriental Robotic Telescopes with a project of a 2-m telescope with around 40 scientists, including M.Sc. and Ph.D. students;

Eastern African countries (such as Ethiopia, Kenya and Madagascar) should carry out analysis of meteorological satellite data and site prospecting investigations on their high mountains in an effort to locate the best observational sites for middle- or large-size telescopes.

The participants noted that regional astronomical newsletters were published and distributed electronically through the Internet and in hard copy on a regular basis, as recommended and supported by the United Nations/ESA workshops since 1996:

> *Africa.* The newsletter *Africa Skies / Cieux Africains* (http://www.saao.ac.za/~wgssa/) was published in a collaborative effort by the South African Astronomical Observatory and the Observatoire Midi-Pyrénées (France);

> *Asia and the Pacific.* The newsletter *Teaching of Astronomy in Asia-Pacific Region* was published by the National Astronomical Observatory of Japan;

> *Latin America and the Caribbean.* The newsletter *Astronomia Latino Americana* (http://www.astro.ugto.mx/~ala/) was published by the University of Guanajuato (Mexico);

> *Western Asia.* Preparations were progressing for the publication of a regional astronomical newsletter under the editorial supervision of an astronomical institute in Saudi Arabia.

2.2.4   Space Technologies

Participants observed that the coordinated effort of the series of workshops on basic space science held under the auspices of the United Nations and ESA served as a catalyst to promote the development of space technology in developing countries, and to make possible cooperative efforts between countries, which would minimize the investment required by individual countries.

Participants noted that the cost of space technology had decreased considerably during the past decades and felt that Governments of developing countries should be encouraged to fund appropriate space science programs for their respective countries in order to enjoy the benefits to be derived from them. They recommended the development of on-line modular courses in space technology at the undergraduate and graduate levels to meet developing countries' needs for basic space science education, preferably in local languages.   They also recommended that developing countries implement curricula to prepare personnel to operate space science programs. Finally, the participants of the workshop recommended considering low-cost nano-satellites as a viable start-up for space projects for developing countries.   Such projects could have a direct impact on the decision-makers of developing countries and could provide a boost for furthering space science research projects.

2.3   Overview of the Series of United Nations/European Space Agency Workshops on Basic Space Science

At the request of the international organizing entities (see §2.2.1), principal national organizers (See Table 4) and participants of United Nations/ESA workshops on basic space science, respectively, information on the workshops held between 1991 and 2000 was gathered for the preparation of an assessment of the achievements of the workshops that could be finalized in 2001-2002.   Subsequently, the results of such an assessment exercise could be brought to the attention of countries interested in the development of space science at the national, regional and international levels.   Tables 1-4 have been compiled by participants of the workshops in cooperation with the principal national organizers of the host countries of all past United Nations/ESA workshops on basic space science.

2.3.1   United Nations/European Space Agency Workshops on Basic Space Science, 1991-2003

Table 1 contains information on host countries, their regional distribution, the number of participants and countries participating in United Nations/ESA workshops on basic space science between 1991 and 2003.   The document symbols of the United Nations reports on the workshops and their titles are also provided.

2.3.2 Regional Distribution and Number of Individuals

The regional distribution of countries or areas and the number of individuals who requested and received information on the results of the United Nations/ESA Workshops on Basic Space Science in 2001 are shown in Table 2.   The addresses of the individuals in their respective countries have been used for the mail and electronic mail (e-mail) distribution of regional astronomical newsletters as described in §2.2.3. The same addresses have been provided to national and international astronomical organizations for dissemination of scientific information.

2.3.3   Projects Undertaken

Projects worked on and follow-up projects pursued through the series of United Nations/ESA Workshops on Basic Space Science from 1991 to 2000 are identified in Table 3. If available, Internet addresses are provided from which detailed information on the respective project can be retrieved. Information on the projects is also contained in the United Nations reports on the workshops listed in Table 1.

2.3.4   Contact Addresses and Published Results

The principal national organizers and participants of workshops have reported, on a continuous basis, on results addressed at and achieved through the workshops. Contact addresses of the principal national organizers at the host institutions can be used to receive updated information on all aspects of the workshops and their results published and reviewed in the international scientific literature. Relevant information is summarized in Table 4.

2.4   Mauritius Radio Telescope

The Mauritius Radio Telescope (MRT) was designed primarily to undertake a survey of the southern sky at 151.5-MHz with a sensitivity of 150- milliJansky. It is also meant to map the Milky Way. A point source catalogue of around 100,000 objects will be produced. The MRT also makes observations of pulsars. Three surveys of the southern sky have already been finalized, with about 300 gigabytes of raw data collected.

The MRT is a synthetic radio telescope that is utilized to take images of the sky at a frequency of 151.5-MHz (or 2-m wavelength). It can detect objects that are too faint to be seen by large optical telescopes. The telescope is a joint project of the Indian Institute of Astrophysics and the Raman Research Institute, both in Bangalore, India, and the University of Mauritius in Reduit. It is located in the Bras d'Eau forest in the rocky north-eastern part of Mauritius (20.14° S and 57.73° E).

The original idea of a southern sky survey at a frequency of 150-MHz was put forward by Ch. V. Sastry of the International Institute of Astrophysics. He had the intention of undertaking a survey equivalent to the Cambridge 6C Survey of the northern sky. Mr. Sastry and V. Radhakrishna of the Raman Research Institute visited Mauritius in 1987 to set up the telescope. Its further development remains a joint project of the three institutions mentioned. The final construction was completed in 1992 and the telescope has been operational since then. W. C. Erikson donated the receiver system from the defunct Clarke Lake Observatory, University of Maryland (United States). The antenna system was designed and built in India.

At present, there are 17 individuals employed for the day-to-day operation of the MRT. They have completed three surveys and a low-resolution map of the southern sky. The processing of data for the final map has been completed. Observations of specific southern sky pulsars were made.

As one result of the establishment and operation of the telescope, four Ph.D.s, one M.Phil. and a number of B.Sc. theses have been completed. Engineers and technicians have been trained in the establishment and operation of the telescope. Subsequently, research papers have been published and, in 1997, the staff organized a conference on low-frequency radio astronomy. The MRT has achieved international recognition.

The MRT is a T-shaped array consisting of 1,020 fixed helical antennas arranged in 32 groups in the east-west arm (2-km long) and 64 helical antennas on 16 movable trolleys in the north-south arm (880-m long). There is a single trolley in the northern arm of the telescope. The antennas collect the radio waves coming from space. The signal from each group is filtered, amplified and sent to the telescope building where it is combined with the signals from other groups. The signal is processed in a correlator and computer programs transform it into images or profiles.

The MRT uses the technique of synthetic aperture to simulate a 1-km by 1-km filled array. Observations are made with the trolleys in the south arm at their nearest position from the center of the array. The trolleys are then moved further south and the observations repeated 62 times. This process continues until the end of the south arm is reached. A computer system, operating with Linux OS, is used to collect the observations to produce a map of the sky. Unlike most radio telescopes, the MRT can see very extended radio sources. Also, the non-coplanarity of the east-west arm has led to the development of new imaging techniques used in cleaning the raw data.

Although the telescope was designed primarily to conduct the 151.5-MHz survey, it has also been used for pulsar observations, for which only the east-west arm is used. The group outputs are added together with a tracking capability of about 20 pulsars for a source transiting at meridian. This corresponds to eight minutes for an equatorial source. Data are recorded at a fast rate over a bandwidth of 1-MHz. The data processing is done to produce an output in the desired format, including the profile unique to each pulsar.

Table 1
United Nations/European Space Agency workshops on basic space science, 1991-2003

Year	City	Target region	Host institution	Number of participants	Number of participating countries	Workshop title	Report
1991	Bangalore (India)	Asia and the Pacific	Indian Space Research Organization	87	19	Basic space science	A/AC.105/489
1992	San José and Bogotá	Latin America and the Caribbean	University of Costa Rica and University of the Andes	122	19	Basic space science	A/AC.105/530
1993	Lagos	Africa	University of Nigeria and Obafemi Awolowo University	54	15	Basic space science	A/AC.105/560/Add.1
1994	Cairo	Western Asia	National Research Institute of Astronomy and Geophysics	95	22	Basic space science	A/AC.105/580
1995	Colombo	Asia and the Pacific	Arthur C. Clarke Institute for Modern Technologies	74	25	Basic space science: from small telescopes to space missions	A/AC.105/640
1996	Bonn	Europe	Max Planck Institute for Radioastronomy	120	34	Basic space science: ground-based and space-borne astronomy	A/AC.105/657
1997	Tegucigalpa	Latin America and the Caribbean	Universidad Nacional Autónoma de Honduras	75	28	Basic space science: small astronomical telescopes and satellites in education and research	A/AC.105/682
1999	Mafraq (Jordan)	Western Asia	Al al-Bayt University	95	35	Basic space science: scientific exploration from space	A/AC.105/723

Year	City	Target region	Host institution	Number of participants	Number of participating countries	Workshop title	Report
1999	Vienna	All regions	United Nations Office at Vienna			(a) Third United Nations Conference on the Exploration and Peaceful Uses of Outer Space (UNISPACE III);	United Nations publication, Sales No. E.00.I.3
						(b) International Astronomical Union (IAU)/Committee on Space Research (COSPAR)/United Nations Special Environmental Symposium "Preserving the Astronomical Sky";	Ibid., annex III, sect. II
						(c) IAU/COSPAR/United Nations Special Workshop on Education in Astronomy and Basic Space Science	Ibid., annex III, sect. VIII
2000	Toulouse (France)	Europe	Centre national d'études spatiales	80	34	Basic space science: satellites and networks of telescopes; tools for global participation in the study of the universe	A/AC.105/742
2001	Reduit (Mauritius)	Africa	University of Mauritius	65	28	Basic space science: exploring the universe; sky surveys, space exploration and space technologies	A/AC.105/766
2002	Cordoba (Argentina)	Latin America and the Caribbean	Institute for Higher Space Studies "J. Mario Gulich" and Universidad de La Plata			Basic space science: World Space Observatory and virtual observatories in the era of 10m telescopes	
2003	Beijing (China)	Asia and the Pacific	China National Space Agency				

Table 2
Regional distribution of countries or areas and number of individuals who requested and received information on the results of the United Nations/European Space Agency workshops on basic space science in 2001

Africa		Asia and the Pacific		Eastern Europe		Latin America and the Caribbean		Western Europe and other States	
Algeria	31	Bahrain	1	Bulgaria	1	Argentina	6	Australia	4
Angola	1	Bangladesh	1	Croatia	1	Bolivia	1	Austria	6
Botswana	3	Brunei Darussalam	1	Czech Republic	1	Brazil	3	Belgium	7
Burkina Faso	1	China	13	Hungary	13	Chile	3	Canada	11
Burundi	2	India	38	Lithuania	38	Colombia	2	Denmark	3
Cameroon	6	Indonesia	8	Poland	8	Costa Rica	7	France	38
Central African Republic	1	Iran (Islamic Republic of)	2	Romania	2	Cuba	5	Germany	49
Côte d'Ivoire	3	Iraq	2	Russian Federation	2	Ecuador	2	Greece	5
Egypt	45	Japan	13	Slovakia	2	El Salvador	6	Ireland	1
Eritrea	1	Jordan	14	The former Yugoslav Republic of Macedonia	13	Guatemala	4	Israel	6
Ethiopia	3	Kazakhstan	3	Ukraine	14	Honduras	24	Italy	15
Gabon	1	Kuwait	9			Mexico	13	Malta	1
Ghana	10	Lebanon	5			Nicaragua	4	Netherlands	4
Guinea	4	Malaysia	2			Panama	3	New Zealand	1
Kenya	12	Mongolia	5			Paraguay	1	Norway	1
Liberia	1	Oman	4			Peru	4	Portugal	2
Libyan Arab Jamahiriya	11	Pakistan	7			Uruguay	6	Spain	14
Madagascar	4	Palestine	1			Venezuela	2	Sweden	3
Malawi	4	Papua New Guinea	3					Switzerland	3
Mali	1	Philippines	3					Turkey	8
Mauritania	3	Qatar	5					United Kingdom of Great Britain and Northern Ireland	15
Mauritius	4	Saudi Arabia	12					United States of America	110
Morocco	23	Singapore	2						
Mozambique	5								
Namibia	4								

Africa		Asia and the Pacific		Eastern Europe	Latin America and the Caribbean	Western Europe and other States
Niger	2	Sri Lanka	6			
Nigeria	77	Syrian Arab Republic	6			
Rwanda	1	Taiwan Province of China	3			
Senegal	2	Tajikistan	1			
Sierra Leone	2	Thailand	4			
South Africa	113	United Arab Emirates	2			
Sudan	4	Uzbekistan	1			
Swaziland	2	Viet Nam	4			
Togo	1	Yemen	2			
Tunisia	8					
Uganda	3					
United Republic of Tanzania	5					
Zaire	2					
Zambia	8					
Zimbabwe	11					

Total number of countries: 124
Total number of individuals: 1,024

Table 3
Projects pursued through the United Nations/European Space Agency workshops on basic space science, 1991-2000

Year	Country	World Wide Web site	Projects worked on at the workshop	Recommended follow-up projects
1991	India		Telescope donation programme of the Government of Japan: Sri Lanka 1995, Paraguay 1999 and the Philippines 2000	Establishment of an astronomical facility at the Arthur C. Clarke Institute for Modern Technologies (ACCIMT) in Sri Lanka
1992	Costa Rica and Colombia	Galactic emission map (Colombia): http://aether.lbl.gov/www/projects/GEM/	Education and career development in basic space science	Establishment of an astronomical observatory for Central America in Honduras
			"ISY92: Planetarium; A Challenge for Educators"	Donation of computer equipment by the European Space Agency (ESA): Cuba, Ghana, Honduras, Nigeria, Peru and Sri Lanka
				Establishment of a 5.5-m radio telescope in Colombia
1993	Nigeria	Inter-African astronomical observatory and science park (Namibia): http://home.t-online.de/home/a.masche/ and http://www.mpia-hd.mpg.de/Public/PUBREL/booklet01.html	Southern African Large Telescope (South Africa)	Establishment of an inter-African astronomical observatory and science park on the Gamsberg in Namibia
		Southern African Large Telescope (South Africa): http://www.salt.ac.za		
1994	Egypt	Kottamia Telescope (Egypt): http://www.sti.sci.eg/scrci/nriag.html	Kottamia Telescope (Egypt)	Refurbishment of the Kottamia Telescope
			Egyptian drill project for the Mars mission	Participation of Egypt in the Russian Federation/United States of America Mars mission 2001
1995	Sri Lanka	ACCIMT telescope facility (Sri Lanka): http://www/slt/lk/accimt/	Inauguration of a telescope facility (Sri Lanka)	Evaluation of the feasibility of a world space observatory
			World space observatory (WSO/UV)	
1996	Germany	Working Group on Space Science in Africa: http://www.saao.ac.za/~wgssa/	Third United Nations Conference on the Exploration and Peaceful Uses of Outer Space (UNISPACE III)	Establishment of NORT
		Network of Oriental Robotic Telescopes (NORT): http://www.saao.ac.za/~wgssa/as2/nort.html	Assessment of the achievements of the United Nations/ESA workshops	

Year	Country	World Wide Web site	Projects worked on at the workshop	Recommended follow-up projects
		Pierre Auger cosmic ray project: http://www.taridar.cnea.gov.ar/~auger/	Foundation of the Working Group on Space Science in Africa	
			NORT	
			100-metre Effelsberg radio telescope	
			Education and research using small astronomical telescopes	
			Developing astronomy and space science worldwide	
			Two air shower detectors, one situated in the northern hemisphere (United States of America) and one in the southern hemisphere (Argentina)	
1997	Honduras	Observatorio Centroamericano de Suyapa (Honduras): http://www.unah.hn	UNISPACE III	Joint membership of Central American countries in the International Astronomical Union
		Space Guard Foundation (Italy): http://spaceguard.ias.rm.cnr.it/	First issue of the newsletter *African Skies/Cieux Africains* published	
			Inauguration of the Central American Astronomical Observatory in Honduras	
			NORT	
			Observation of near-Earth objects	
1999	Jordan	Maragha Astronomical Observatory (Jordan): http://www.aabu.edu.jo/	UNISPACE III	Operation of the astronomical telescope facility at Al al-Bayt University
		Hands-on astrophysics: http://www.aavso.org/	WSO/UV	Planning of the 31-m Baquaa radio telescope at the University of Jordan
		Astrophysics education module for university physics courses: http://www.seas.columbia.edu/~ah297/um-esa/astrophysics/index.html	Operation of the Maragha Astronomical Observatory in Jordan	
			Baquaa radio telescope	
			Hands-on astrophysics	
			Astrophysics for university physics courses	
2000	France	World space observatory/UV: http://www.seas.columbia.edu/~ah297/um-esa/wso.html	UNISPACE III	WSO/UV assessment study completed
			WSO/UV	
			NORT	
			Regional newsletters on astronomy	

Table 4
Contact addresses and published results of the United Nations/European Space Agency workshops on basic space science, 1991–2001

Year	Principal organizer	Published review of workshop	Working papers published in Seminars of the United Nations Programme on Space Applications: Selected Papers from Activities	Workshop proceedings
1991	S. C. Chakravarty Indian Space Research Organization Antariksh Bhavan New BEL Road Bangalore 560 094 India scc@isro.ernet.in	*Astrophysics and Space Science*, vol. 193, 1992, p. 161.	One working paper each in Nos. 3 (1992) and 4 (1993)	*AIP Conference Proceedings*, vol. 245, 1992, pp. 1-350.
1992	Walter Fernandez School of Physics University of Costa Rica 2060 San José Costa Rica wfer@cariari.ucr.ac.cr	*Earth Space Review*, vol. 2, No. 2, 1993, pp. 25 and 26. *COSPAR Information Bulletin*, vol. 2000, No. 149, pp. 82-84.	No working papers published	*Earth, Moon, and Planets*, vol. 63, No. 2, 1993, pp. 93-179.
1992	Sergio Torres Observatoria Astronómico Nacional Universidad Nacional de Colombia P. O. Box 2584 Santa Fe de Bogotá Colombia verada@earthlink.net	*Earth Space Review*, vol. 2, No. 2, 1993, pp. 25 and 26. *COSPAR Information Bulletin*, vol. 1999, No. 144, pp. 13-15.	No working papers published	*Astrophysics and Space Science*, vol. 214, 1994, pp. 1-260.
1993	Pius N. Okeke Space Research Centre University of Nigeria Nsukka Nigeria misunn@aol.com	*Earth Space Review*, vol. 3, No. 3, 1994, pp. 26 and 27. *COSPAR Information Bulletin*, vol. 1999, No. 144, pp. 28-30.	Three working papers in No. 5 (1994)	*AIP Conference Proceedings*, vol. 320, 1995, pp. 1-320.

Year	Principal organizer	Published review of workshop	Working papers published in Seminars of the United Nations Programme on Space Applications: Selected Papers from Activities	Workshop proceedings
1994	Joseph S. Mikhail National Research Institute of Astronomy and Geophysics Helwan Cairo Egypt	*Earth Space Review*, vol. 4, No. 2, 1994, pp. 28-30. *COSPAR Information Bulletin*, vol. 2000, No. 148, pp. 41 and 42.	Three working papers in No. 6 (1995)	*Earth, Moon, and Planets*, vol. 70, Nos. 1-3, 1995, pp. 1-233. *Astrophysics and Space Science*, vol. 228, 1995, pp. 1-405.
1995	Padmasiri De Alwis Arthur C. Clarke Institute for Modern Technologies Katubedda, Moratuwa Sri Lanka <u>asela@slt.lk</u>	*COSPAR Information Bulletin*, vol. 1996, No. 136, pp. 8-11. *ESA Bulletin*, No. 81, February 1995, pp. 18-21.	Three working papers in No. 8 (1997)	
1996	Rolf Schwartz Max Planck Institute for Radioastronomy Auf dem Hügel 69 D-53121 Bonn Germany rolf@mpifr-bonn.mpg.de	*COSPAR Information Bulletin*, vol. 1997, No. 138, pp. 21-24. *AAS Newsletter*, No. 79, 1996, pp. 18 and 19.	Two working papers in No. 8 (1997)	*Astrophysics and Space Science*, vol. 258, 1998, pp. 1-394.
1997	Maria Cristina Pineda de Carias Observatorio Astronómico Nacional Autónoma de Honduras Apartado Postal 4432 Tegucigalpa M.D.C. Honduras <u>Mcarias@hondutel.hn</u>	*COSPAR Information Bulletin*, vol. 1998, Issue 141, pp. 9 and 10. *Annals of the New York Academy of Sciences*, vol. 822, 1997, pp. 621-630.	Six working papers in No. 9 (1998)	
1999	Hamid M. K. Al-Naimiy Higher Institute of Astronomy and Space Sciences Al al-Bayt University P. O. Box 130302 Al Mafraq Jordan <u>Alnaimiy@yahoo.com</u>	*COSPAR Information Bulletin*, vol. 1999, Issue 146, pp. 9 and 10.	Six working papers in No. 11 (1999)	*Astrophysics and Space Science*, vol. 273, 2000, pp. 1-343.

Year	Principal organizer	Published review of workshop	Working papers published in Seminars of the United Nations Programme on Space Applications: Selected Papers from Activities	Workshop proceedings
2000	François R. Querci Observatoire Midi-Pyrénées 14, avenue Édouard Belin F-31400 Toulouse France fquerci@ast.obs-mip.fr	*COSPAR Information Bulletin*, vol. 2000, No. 149, pp. 66 and 67. *AAS Newsletter*, No. 100, June 2000, p. 21. *AAS Newsletter*, No. 102, October 2000, p. 14.	Thirteen working papers in No. 12 (2001)	
2001	Soonil D. Rughooputh Faculty of Science University of Mauritius Reduit Mauritius soonil@uom.ac.mu	*COSPAR Information Bulletin*, vol. 2001, No. 152, pp. 16-18. *AAS Newsletter*, No. 107, October 2001, pp. 16 and 17.	Twelfe working papers in No. 13 (2002)	Astrophysics and Space Science (in print)

3. REFERENCES

Bennett, J. 2001, "On the Cosmic Horizon: Ten Great Mysteries for Third Millennium Astronomy", Addison Wesley Longman, San Francisco, California.

Bennett, J. Donahue, M., Schneider, N., and Voit, M. 2002, "The Cosmic Perspective", Second Edition, Addison Wesley, San Francisco, California.

Burdyuzha, V. and Khozin, G. (eds.) 2000, "The Future of the Universe and the Future of Our Civilization", World Scientific, Singapore.

Heck, A. (ed.) 2001, "Organizations and Strategies in Astronomy II", Kluwer Academic Publishers, Dordrecht 2001.

Kitamura, M. 1999a, "Provision of astronomical instruments to developing countries by Japanese ODA with emphasis on research observations by the donated 45-cm reflectors in Asia", in Conference on Space Sciences and Technology Applications for National Development: Proceedings, held at Colombo, Sri Lanka, 21-22 January 1999, Ministry of Science and Technology of Sri Lanka.

Kitamura, M. 1999b. "Provision of astronomical equipment for developing countries through ODA of the Government of Japan", Seminars of the UN Programme on Space Applications: Selected Papers from Activities Held in 1999, No. 11, UN Document ST/SPACE/3.

National Academy Press 2001a, "Astronomy and Astrophysics in the New Millennium: Panel Reports", National Academy Press, Washington, D.C.

National Academy Press 2001b, "Astronomy and Astrophysics in the New Millennium", National Academy Press, Washington, D.C.

Oswalt, T.D. 1998, "Preserving Access to Small Observatories Throughout the World: The Role of University Consortia and Collaboration", I.A.P.P.P. Communications No. 70, winter 1998.

Wentzel, D.G. 1998, Astrophysics for University Physics Courses, College Park, Maryland, University of Maryland, 1998 (English version available at:

http://www.seas.columbia.edu/~ah297/un-esa/astrophysics/.

Chapter 19

Important Roles of Small Telescopes in Space

Yoji Kondo
NASA Goddard Space Flight Center
Greenbelt, Maryland USA

Abstract: Small telescopes play a large role in astronomy today. Whether in space or on Earth, they still produce outstanding results.

Key words: telescopes, small; telescopes, space-based

1.   INTRODUCTION

This article is being written in part to dispel a popular misconception about the roles played today by small telescopes in astronomy. Small telescopes do play indispensable roles in pushing the frontier of research, but the funding situation for small telescopes has become increasingly difficult in recent years with the ever-increasing number of big telescopes competing for funds from basically the same pot of money. When there is a contest of this nature, bigger telescopes, whose proponents have greater political muscles than those supporting small telescopes, usually win out.

Do not misunderstand me. I am not at all saying that we should not have those great telescopes on the ground and in space; in fact, I believe that these large telescopes perform wonderful roles in astronomy and are essential in advancing our research. Here, I am simply trying to point out the continuing importance of small telescopes in modern astronomical research.

The arguments not to fund small telescopes often go as follows: big telescopes can do anything that small telescopes are capable of performing; therefore, there is no need to build or keep operating small telescopes once big telescopes are in place. They might maintain, in addition, that the most exciting discoveries made today are from big telescopes that are primarily used in cosmological and extragalactic research. The antagonists (to small telescopes) might even go further. Why waste our time studying stars? We

T.D. Oswalt (ed.), The Future of Small Telescopes in the New Millennium, Vol. I, 227–237.

already know just about all we need to know about stars, except, possibly, filling out some relatively minor details about their properties.

Such a view is entirely mistaken, both scientifically and financially. Those twin aspects of research—scientific and financial ones—are closely connected to each other and cannot be treated as separate issues. In consequence, we need both big and small telescopes in astronomy.

It might also be noted that a number of small telescopes in operation would also mean that there would be more—and different—people making decisions on telescope time allocation. Since we cannot predict the outcome of new ideas in advance, this would tend to help innovative research projects that might otherwise be considered too risky.

Understanding the physics of stars is fundamental to practically all other aspects of astrophysics. Studying the universe without understanding the stars (which are their building blocks) is somewhat analogous to conducting research in modern physics without understanding atomic and subatomic physics. In order to understand the galaxies and the cosmos, we must first understand what makes up galaxies, i.e., the stars and the matter between the stars or the interstellar gas and dust.

A 10-m telescope, which is currently the largest telescope on the ground, would not be used to study the light variations of δ Cephei or RR Lyrae in distant galaxies, had they not already been known as important distance indicators through their earlier observations in the Magellanic Clouds. Such a large telescope would be too expensive to be used for conducting research that could be pursued efficiently with significantly less costly smaller telescopes.

Those intrinsic variable stars remain two of the important standard candles used in measuring cosmic distances. Large telescopes are used today to observe the light variations of δ Cephei type variables in distant galaxies only because their importance as standard candles are known through earlier observations using smaller telescopes.

Furthermore, in order to use Cepheids as reliable distance indicators, we have seen that it is important to study them in detail within our own and in nearby galaxies. Since the discovery of their importance as standard candles, we have learned that there are two kinds of Cepheids, Types I and II, with different period absolute luminosity relationships. Without such insights, our cosmological research could have gone astray through the misjudging of cosmological distances.

And, what about the physical properties of Type Ia supernovae? They are used more and more as standard candles to determine distances to distant galaxies because of their great intrinsic brightness; we need to study Type Ia supernovae in detail and in sufficient numbers to make certain of our suppositions about their properties.

Who knows what new surprises might be in store as we learn more about the stars around us?

Let us now examine the financial issues involved. For one thing, large telescopes cost significantly more than small telescopes to build and operate. We must be mindful of the limitation in our resources, since our funds for building and operating various telescopes are not unlimited. The costs of building and operating large telescopes with an aperture of, say, 10-m is more than an order of magnitude greater than those for smaller telescopes with an aperture of around 1-m or less. While it is possible in principle to use a 10-m aperture telescope to obtain the spectra of nearby stars, it is unlikely that there will be enough such large telescopes to be used for such purposes.

Consider the following metaphor: one does not drive an 18-wheeler to pick up a carton of milk from the corner grocery store; you drive your family car or even walk there for such an errand. There are different vehicles that are suited for diverse purposes in life, economically speaking or otherwise—as do telescopes of various sizes for diverse scientific objectives.

For training future astronomers, both undergraduate and graduate students, small telescopes (usually with less stringent requirements for making every minute of telescope time to count toward publications) are excellent educational laboratories. It allows the students to learn the profession without the severe pressure to make everything work perfectly from the outset—even being permitted to make small (reparable) mistakes in the process.

The financial considerations get even more serious when talking about observatories in orbit. In addition to the cost of building telescopes, we need to consider the cost of placing them in orbit. At the moment, the cost of launching a kilogram of payload into low Earth orbit (LEO) is around $10,000 to $20,000 per kilogram; it would of course cost substantially more to place them in geosynchronous or other more distant orbits.

It is hoped that the development and deployment of economical reusable launch vehicles (RLVs) will reduced the cost of reaching LEO by an order of magnitude at first and, within a few decades, by two orders of magnitude to some $100 per kilogram. At the moment, however, we should not be making near-future plans for observatories in space based on such an anticipation.

At such considerable costs, we need to focus clearly on what we are trying to accomplish with our limited resources and on how to go about it. Some types of research will certainly require large aperture telescopes, such as the Hubble Space Telescope and the Next Generation Space Telescope, whereas some other kinds of research can be performed efficiently with smaller aperture telescopes. Let us first examine how a 45-cm telescope in geosynchronous orbit contributed to astronomical research from space and see how in the future an array of small aperture telescopes—e.g., from 50- to 100-cm aperture—can still make important contributions to our science.

There will be other articles in this volume discussing the importance of small telescopes on the ground.

The examples of outstanding small telescopes that leap to mind immediately include the Fairborn Observatory; the 0.8-m automated photoelectric telescope there is operated by the Four College Observatory Consortium (consisting of the College of Charleston, The Citadel, University of Nevada-Las Vegas and Villanova University); it is located in the Patagonia Mountains in Arizona. There also is a pair of productive 0.8-m telescopes, named appropriately Amadeus and Wolfgang, operated by the University of Vienna. Another excellent example is the CHARA Interferometric Array of several 1-m class telescopes at Mount Wilson, operated by the group headed by Harold McAlister of the Georgia State University. Yet another interesting case is the 2.5-m telescope at Mount Wilson, which aperture, by the definition used in compiling this book, consigns it to the realm of a small telescope. With adaptive optics, the once famous 100-in telescope is again performing important astronomical research.

In this article, we shall therefore focus on small telescopes in space.

2.   A HISTORICAL EXAMPLE OF PRODUCTIVE SMALL TELESCOPES IN SPACE, IUE

Over the past few decades, there were a number of small telescopes that produced good scientific results—without even counting small X-ray, Infrared, and solar observatories. To name a few examples of ultraviolet satellites (which make for a natural extension of ground-based observatories in the sense that observing techniques employed are basically similar to those on the ground) there were the Orbiting Astronomical Observatory (OA0-A2), Copernicus (OAO-3), Astronomical Netherlands Satellite (ANS), and Extreme Ultraviolet Explorer (EUVE). The EUVE, which bridged the wavelength region between the X-ray and ultraviolet, was launched in June 1991 and ended its operation at the end of January 2001. The EUVE observed more than 1230 astronomical sources in our galaxy (of which about a quarter remain unidentified) and some 20 extragalactic objects—quite contrary to the predictions of some astronomers who had thought that the interstellar space would be too opaque to the extreme ultraviolet photons to see beyond a few light years.

In this article, as a good example of small telescopes in space that produced outstanding astronomical results, let us review the 45-cm aperture International Ultraviolet Explorer (IUE) satellite; the IUE was placed in a geosynchronous orbit over the Atlantic in January 1978 and its scientific operation was terminated at the end of September 1996. A collaborative project with the European Space Agency and the British Science and Engineering Council, the IUE was operated 16 hours a day from Goddard Space Flight Center in Maryland and 8 hours a day from the Villafranca ESA ground station, near Madrid Spain. The annual budget for the NASA operation of the IUE, including the grants to approximately one hundred guest observers per year, was in the range of 10 million dollars or less—mostly less—in the U.S.

Operating continuously 24 hours a day and 365-6 days a year, which was enabled by its geosynchronous orbit, the IUE was used to obtain low (300) and high (10,000) resolution spectra in the far-ultraviolet (1150-2000 A) and mid-ultraviolet (1800-3200 A). Over 100,000 low and high resolution spectra of various astronomical sources were obtained and are readily available from the archive to any interested individuals in the world. Those interested may visit the MAST/IUE Final Archive website[1]; its usage is self-explanatory. The list of observed targets include from nearby objects such as planets, planetary satellites, comets and asteroids, to all types of stars, interstellar medium, and a variety of galaxies, including active galactic nuclei and quasars (Kondo et al. 1989; Wamsteker, Kondo & Stickland 1998).

Here briefly are some of the scientific highlights of the IUE observatory:

- Comprehensive studies of stellar winds and mass loss, especially in hot stars and evolved stars.

- Extensive observations of the chromospheres of cool stars.

- Improved understanding of the evolutionary processes in interacting binary stars, which in the past had largely been a matter of speculation based on incomplete data.

- Ditto for the physical processes in cataclysmic binaries.

- Supernovae, including SN1987A where the identification of the progenitor (a historical first) was made possible with the IUE data. Contrary to the prevailing ideas, it turned out to have been a blue supergiant (rather than a red supergiant).

- Observations of the high temperature gas associated with pre-main-sequence stars, such as T Tauri and Herbig-Haro objects.

- Pioneering studies of proto-planetary systems, such as β Pictoris.

- Detection, observations and studies of the galactic halos.

- Significantly improved understanding of the distributions and physical conditions of the local interstellar medium.

- Observations of the Super-Bubbles in the Galaxy.

---

[1] http://www.stsci.edu/cgi-bin/iue

- Extensive studies of the Magellanic Clouds, our two satellite galaxies.

- Synoptic observations of active galactic nuclei; thanks to its 24-hour availability, the IUE served as the cornerstone in most, if not all, of the multi-wavelength observations from X-ray to radio.

- Synoptic studies of solar system objects, such as the Io torus.

- Continuous observations of periodic and new comets, such as Halley's Comet and IRAS-Araki-Alcock. Halley observations, for example, provided crucial data on the composition of the cometary comae and tails. The IRAS-Araki-Alcock comet observations led to the detection of diatomic sulfur (S_2) for the first time in a comet.

- The IUE was able to provide a uniform UV spectral catalogue for a variety of astronomical objects. It also allowed long temporal studies of selected objects with a uniform database.

At the last count as of October 2000, the total of refereed papers based on IUE observations was 3,725; this number is still climbing due to the use of the easy to use archival data. For the actual list of publications in refereed journals, visit the website[2].

The continuous 24-hour operation contributed significantly to this unparalleled productivity. The flexibility of its operation also helped immensely, especially to observe targets of opportunity such as novae and supernovae. When SN1987a brightened in the Large Magellanic Cloud in 1987, the IUE was ready turn to it immediately. (I had previously issued a standing order to the observatory staff to switch to any nearby supernova immediately if one is reported. Bob Kirshner had the target of opportunity program to observe supernovae that year.) The IUE data was a key factor in determining what the progenitor star was for the first time in astronomical history. Contrary to the prevailing theory, the progenitor turned out to have been a blue supergiant (rather than a red supergiant).

The manner of the operation of the IUE observatory—i.e., the relatively low cost of operations and the fact that the observing time was more plentiful due to the 24-hour availability of the observatory—made it possible to give some of its valuable observing time to projects that may normally be considered high-risk projects, that is to say, the programs for which scientific returns were rather uncertain based on the (then) prevailing theories and premises. This too contributed to the success of IUE.

It might be worthwhile to note that, of its life time of 18 years and 8 months, the IUE was operated with only two gyros for about 10 years and a half (more than one half its lifetime) and with only one gyro during the last several months, until its operation was ended by a command from the ground

---

[2] http://ines.vilspa.esa.es/ines/docs/iue_publ.pdf

at the end of September 1997. The original satellite design required a minimum of three gyros to sense the three spatial dimensions; the satellite was equipped with 6 gyros to allow for up to three gyro failures. The IUE had the designed life expectancy of 3 to 5 years. (Most satellite observatories have lost their gyros after a few years in orbit; the IUE was the first astronomical observatory to operate in space with only two gyros.) The extended life of IUE by more than one hundred percent was thanks to the flexible design of the satellite and the ingenuity of the satellite guidance and control engineers at Goddard Space Flight Center and of the staff astronomers who operated the IUE at NASA Goddard and ESA Villafranca ground stations.

The easy access to the IUE to a large number of observers made it an excellent training ground for those who had never used spaceborne telescopes. Indeed a large fraction of those who became Hubble Space Telescope observers in its first several years cut their teeth with the IUE—a number of them as graduate students or as postdocs.

3. AN INTERFEROMETRIC ARRAY OF 1-M CLASS TELESCOPES ON THE MOON: AN EXAMPLE OF UNIQUE SMALL SPACE TELESCOPES OF THE FUTURE

Rather than enumerating several examples of small telescopes in space that can contribute significantly to astronomical research in various fields, we will examine one case that can provide unique astronomical data that cannot be obtained any other way with the existing technology.

First, because of its geological inactivity, the Moon is an immense and stable optical bench in the vacuum of space. With optical cables or transmission towers, a base line of tens of kilometers—even hundreds of kilometers, in principle—is an achievable goal with the application of the currently understood technology. The first interferometric array of one-meter class telescopes could be set up there at a cost in the range of NASA's Mid-Ex programs, i.e., a few hundred million dollars that are at least comparable, if not less costly than, other interferometric missions in space (Kondo et al., 2000), such as the Terrestrial Planetary Finder Mission (Stone, 2001).

Other interferometric missions being planned, such as the Space Interferometry Mission (SIM) and Deep Space 3 missions, would be valuable in testing new technologies. The SIM mission will consist of several small aperture telescopes, which are to be mounted on an orbiting optical bench about 10-m long. If Deep Space 3 mission is successful in demonstrating that the relative positions of three orbiting telescopes can be determined continuously with a precision necessary for an interferometric array of free orbiting telescopes, it would be a great boon to astronomy. The precision required is in the range measured in angstroms; such a technology is yet to be demonstrated.   (An important ground-based interferometric array, the CHARA, at Mount Wilson was mentioned earlier.)

In the meanwhile, the Moon can serve as a huge optical bench in vacuum, free of any atmospheric distortions of images. What makes an interferometric array on the Moon feasible in terms of its engineering and costs?

First, Peter Chen of the Catholic University of America and NASA Goddard Space Flight Center has succeeded in developing extreme lightweight and sturdy diffraction-limited optics. For example, the areal density of the mirror used in the Hubble Space Telescope is about 180 kg m^{-2}. In contrast, the lightweight optics developed by Peter Chen has an aerial density of 6 kg per square meter. He soon expects to develop a mirror with an areal density of 1 kg m^{-2} (see Chen & Romeo, 2001; Chen et al., 2000, Chen, Oliversen, Kondo, 2000)!

Other innovations to aid in placing an interferometric array on the Moon include the development of Charge Injection Detectors (CIDs), which are silicon-based solid state array sensors that can withstand solar flares for decades with minimum shielding. We can also use a low-cost, agile missile interceptor known as the Light Exo-Atmospheric Projectile (LEAP) as a small payload lunar lander at a modest cost of less than $100,000 a piece.

The most cost effective lunar interferometer in the near future would probably be a dilute synthetic aperture optical interferometer with a one-kilometer baseline (a limit imposed by the natural curvature of the Moon). The baseline corresponds to a resolution of 0.1 milliarcsecond in visible light, or more than a few hundred times that of the HST. Using a suitable crater or a valley this baseline could be extended significantly without using optical cables or transmission towers. In the future, the use of advance technologies, such as optical cables, could extend the baseline to many times that of the initial observatory array. The important point here is that we already have the necessary technology to deploy six 1-m class telescopes (i.e. with a collecting area of 1-m^2) on the Moon in an interferometric array, in contrast to worthy but currently unavailable technology such as that being tested in Deep Space 3. The light-gathering capability of an array will go up as a multiple of the telescopes used; the effective light-collecting area in this instance will be $1 \times 6\text{-m}^2 = 6\text{-m}^2$.

Given a certain light-collecting area, the depth of cosmos that you will see depends on the resolution of the optical instrument. (The higher the resolution is, the higher the signal-to-noise ratio will be for the target.) Thus, the higher resolution thus achievable using an array of interferometric telescopes on the Moon can provide answers to many important unsolved issues in stellar astronomy, galactic and extragalactic research, and cosmology.

To extra-solar-system planet hunters: on the Moon using optical cables or relay towers, it is in principle feasible to have a base-line for interferometry of up to 1,000 km. If you are observing at the wavelength of 1 micron, a putative solar system 10 light years (some 10^{19} cm) away, you will have a resolving power of 1,000 km at the distance of the target. If you are observing at 10 microns, at which an Earth-like planet will probably be easiest to image, the image resolution will decrease to 10,000 km. On the other hand, apodizing and inteferometric technologies have made significant progresses; it is now possible to block out most (nearly all) of the light from the primary star in that putative solar system. Imaging an Earth-like planet 10 light years away need not be an unachievable dream.

To cosmologists: There may be a practical limit to the size of telescope. However, in an interferometric array, it is possible to use multiple telescopes to gather more photons. Coupled with the much higher image resolution achievable, this means that we can push the frontier of astronomy further back—literally. If we have the capability of looking further than, say, 15 billion light years clearly, we can obtain data to address the question of whether the boundary of the universe, if there be such, is restricted to the neighborhood of 13-15 billion light years in accordance with the current Inflationary Big Bang cosmological model. If we see nothing beyond 15 billion light years, in spite of the telescopes' capabilities to see further, that would provide an argument in favor of the universe having started expansion (from an extremely small volume) some 15 billion years ago. Of course, in order to pursue such a scientific objective meaningfully, we will have to develop a reliable method for determining such great distances, quite independent of the cosmological model being investigated.

On the other hand, if we see astronomical sources much further than 15 billion light years, it would be very, very interesting—to say the least. After all, astronomy is basically an observational science. Tycho Brahe's precision observations of planets, especially those of Mars, started the Keplerian revolution in astronomy—which revolution Tycho had no intention of initiating.

4. MORE FUTURE SMALL OBSERVATORIES IN SPACE

Examples of small orbiting telescopes in various planning stages are cited below. (Note: this list is not intended to be all-inclusive.)

4.1   Kepler[3]

Kepler is a mission designed to detect Earth-sized planets in habitable zones and also to monitor over 100,000 sun-like stars and to monitor up to 42,000 variable stars continuously for a period of up to 4 years. It has been approved for launch in 2007.

4.2   Eddington[4]

The mission objectives are to understand the physical processes that govern the evolution of stars of different types and ages, and to search for and determining the characteristics of Earth-like planets orbiting those stars.

4.3   COROT[5]

COROT is dedicated to stellar seismology and the study of extra-solar-system planets.

4.4   MOST[6]

MOST, weighing only 50-kg, will measure the oscillation in light intensity of stars in order to determine their composition as well as ages.

5.   ACKNOWLEDGEMENTS

I would like to thank Frederick C. Bruhweiler of the Catholic University of America, Edward F. Guinan of Villanova University, and George E. McCluskey of Lehigh University for helpful discussions in writing this article.

6.   REFERENCES

Chen, P.C. & Romeo, R.C., Laser Focus World, 193, 2001.
Chen, P.C., Bowers, C.W., Conent, D.A., Marzouk, M. & Romeo, R.C., Opt. Eng., 2320, 2000.
Chen, P.C., Oliversen, R.J. & Kondo, Y., "A First Generation Lunar UV Observatory", SPACE 2000, Proc. Space 2000: The 7th International Conference and Exposition on

---

[3] www.kepler.arc.nasa.gov
[4] Those interested may use a search engine for "Eddington ESA" to find the current site.
[5] Use a search engine for "COROT French Space Agency"
[6] Use a Search Engine for "MOST Canadian Space Agency"

Engineering, Construction, Operations, and Business in Space, eds. S.W. Johnson, K.M. Chua, R. Galloway & P. Richter, Am. Soc. Civil Eng., 712-718, 2000.

Kondo, Y., Wamsteker, W., Boggess, A., Grewing, M., de Jager, C., Lane, A.L., Linsky, J.L. & Wilson, R., (Editors), "Exploring the Universe with the IUE Satellite", Kluwer Academic Publishers, 1989.

Kondo, Y., Terzian, Y., Chen, P., Mendell, W.W., Oliversen, R.J., Artemis, 14 (Summer 2000).

Stone, E., in "Space Access & Utilization Beyond 2000", eds. Y. Kondo, C. Sheffield, F. Bruhweiler, Amer. Astronautical Society: Science & Technology Series, Volume 101, p.93, 2001.

Wamsteker, W., Kondo, Y. & Stickland, D., (Organizers), Symposium on "Ultraviolet Astrophysics Beyond the IUE Final Archive", ESA SP-413 (1998).

Chapter 20

Telescope Performance: Past, Present, and Future

Frank Melsheimer
DFM Engineering, Inc.
Longmont, Colorado USA

Abstract: Historically telescopes and instruments have been controlled by the observer at the guiding eyepiece in the cold and dark observatory. Here the observer performed boring menial tasks for hours to capture a small amount of astronomical data. Today most observing is being performed from an air-conditioned "warm" room using a very efficient CCD detector. The observer is assisted by a computer that performs the menial tasks at a level not even dreamed possible 30 years ago. In a few years, the observer will be partially replaced with sophisticated computer software that will schedule the observations, control the telescope, operate the instruments, gather the data, preprocess the data, and transmit the data to the "observer". The past, present, and future performance requirements for telescopes are examined in terms of optics, pointing, tracking, controls, structures, and support functions.

Keywords: telescopes: automated, remote operation, efficiency

1. PAST GENERATION TELESCOPE PERFORMANCE

1.1 Optics

Thirty years ago I became intimately familiar with telescopes and optics ranging in aperture from 8-in up to the 120-in (3-m) Shane telescope at Lick Observatory's Mt. Hamilton. My job at Lick was to make the telescopes more efficient although the word "efficient" was never used. I found that the telescopes were perceived to have adequate to good optics in that the optics were nearly always seeing limited. Optical testing was primarily done with a knife-edge and in a few cases a Hartman screen. During the next five years,

T.D. Oswalt (ed.), The Future of Small Telescopes in the New Millennium, Vol. I, 239–256.
© 2003 *Kluwer Academic Publishers. Printed in the Netherlands.*

we completed the Cassegrain secondary for the 3-m, a new primary mirror for the Crossley telescope, and camera optics for several spectrographs. Both telescope projects were tested using the classical knife-edge test. The 3-m secondary was tested in the telescope using stars and the Crossley mirror was tested in autocollimation in the optical shop. The figuring was judged to be completed based upon the knife-edge test results and "tribal knowledge". The actual optical quality is not quantitatively known.

Collimation of the optics was performed by examining the images in an eyepiece and mechanically adjusting the optics for the best image quality. This involved tip-tilt and centering adjustments of the primary and secondary mirrors. When the seeing is good, these adjustments lead to acceptable images. Unfortunately, when the seeing was at its best, no observer would give up their observing time to allow the telescope optical collimation to be improved, so the image quality was never as good as possible.

Typically focus was adjusted by observing the image reflected off of the slit jaws or with the guiding eyepiece. Photographically, a series of exposures was made and the best focus determined. Sometimes a graph of focus versus temperature was produced to allow the observer to change the focus during the night to correct for temperature changes. Fortunately, at many of the good astronomical sites, the temperature is nearly constant all night minimizing the focus change with temperature change. Using an eyepiece, the focus could be set fairly quickly. Photographically, the determination of best focus was very time consuming.

Most mirror substrates were Pyrex. A near zero thermal expansion glass-ceramic material was introduced (Cervit) and later two other near zero expansion materials became available: ZeroDur from Schott Glass Works and Ultra Low Expansion (ULE) from Corning Glass Works. A borosilicate glass similar to Pyrex is still being used in the form of a spin cast lightweighted mirror substrate and in a fabricated honeycomb substrate.

The telescope effective focal length was relatively long as the telescopes were designed to support photometry and spectroscopy. Imaging was done most often using photographic plates that came in various sizes. The smallest photographic plates were usually larger than present day large CCDs.

1.2   Detectors

The standard photographic plate with its very large format (perhaps 400 million pixels) was the primary detector. The limited linearity and the shallow depth (perhaps 8 bits) were the major limitation of the photographic plate. Electronic detectors were beginning to be used, first the photomultiplier tube, and then several forms of area detectors using cathodes such as the Kron tube, image intensifiers, image dissectors, cooled and integrating vidicons.

1.3   Pointing

The pointing of the telescope was performed by the observer (or night assistant) by using setting circles or equivalent. Good tribal knowledge of the telescope and estimating refraction corrections would allow a skilled observer to find the next desired object in a few minutes. Putting a value on the pointing accuracy is difficult, but at best, the typical pointing would be rated at 10- to 20-arcmin r.m.s. The telescope was slewed by pressing buttons on a hand paddle. Typically, the final part of the slew was performed one axis at a time. This considerably increased the time to move the telescope from one object to another.

1.4   Tracking

The older generation of telescopes in general used very finely made worm gear drives. These drives produced periodic tracking errors often as good as 1-arcsec peak-to-valley and had a period of typically 2- to 4-minutes. The open loop tracking (no optical feedback) was more limited by the proper velocity commanded than by the gearing. Even short exposures (integrations in today's nomenclature) required guiding.

1.5   Controls

The control systems were designed to provide the track, guide, set, and slew speeds using primarily AC motors. Because of the very limited dynamic velocity range of the motors, the four speeds were provided by four separate motors. The velocities were summed using mechanical differentials and switching between motors with electro-mechanical clutches. The speeds were selected by pushing the appropriate buttons on the hand paddle.

The mechanical complexity of the drive system was high with many areas where lost motion (backlash) could be introduced. The electrical complexity was high with many relays performing the logic and commanding the motors. The resulting hardware was expensive and required a fair amount of maintenance.

Worm gear drive efficiency is low which prevents the gearing from back driving (acting as a speed increaser). This makes removing kinetic energy from the telescope difficult. The telescope must be slowly decelerated to tracking speed. A high inertia at the motor is needed to protect the gearing. These factors increase the time required to move the telescope from object to object.

1.6   Structures

A significant advance in the design of telescope structures occurred during the past generation. The 200-in Hale telescope at Mt. Palomar was a great achievement involving structural analysis just barely possible with the tools and knowledge available. The inventions of the Serrier truss and hydrostatic bearings made possible this telescope and these developments are still used today.

The movement to fork mounts was begun with the Palomar 1.2-m Schmidt camera, and the Lick 3-m. These early large fork-mounted telescopes provided some efficiency because of their symmetry. They used less counterweight and did not have to be reversed to the other side of the pier like unsymmetrical mounts. Structurally, the fork geometry produced some non-symmetrical flexures in R.A. that affected the pointing. For example the Lick 3-m has 4 arc minutes of differential fork flexure. This affects the pointing and the tracking of the telescope.

In 1972 I introduced the idea that telescopes should have a high natural resonant frequency. This means high stiffness and low inertia. Such a system would respond to positioning commands with less lost time and would also have less wind shake. The typical telescope, including the Lick 3-m, has a natural resonant frequency in the 0.5- to 2-Hz range. This highly limited the telescope's ability to be moved rapidly even over small distances. The 3-m telescope was using a two-slit spectrograph. Every few minutes, the telescope was moved about 40 arcsec to place the object on the other slit to average out the instrumental differences. With a faint object, the routine movement could take a significant fraction of the integration time. By computerizing the motion, we were able to provide the motion in only a few seconds thus increasing the efficiency of the data collection.

1.7   Support Functions

Automatic dome control was provided on the Palomar and Lick telescopes, but was not wide spread. Typically, the dome was operated by the observer. The concept of "dome seeing" was just becoming recognized as important.

2.   PRESENT GENERATION TELESCOPE PERFOMANCE

The development of the Charge Coupled Device (CCD) has driven telescope performance requirements very rapidly. While the quantum efficiency and deep well capability of the CCD have allowed detecting faint objects well beyond photographic plates, the small format of the detector and its very efficiency have put demands upon the telescope not previously required. When an exposure required several hours, several minutes spent moving to the next object was not important. Now that same limiting image may be obtained in less than a minute and several images are required to cover the same field of view—moving to the next field cannot take several minutes.

Focusing of the telescope must also be made more efficient and needs to be insensitive to temperature changes. The CCD reads out rather slowly, so the focus position mechanism needs to be very repeatable and quickly adjusted. Focus change caused by temperature change and perhaps zenith distance must also be compensated.

2.1   Optics

Presently telescope optics are tested interferometrically. The average optical quality has improved substantially to where 1/4 wave peak-to-valley and 1/10 wave r.m.s. measured on the wavefront at 633-nm are considered to be standard values. The primary focal ratio for Cassegrain telescopes has become much faster. Where $f/3.3$ for the Palomar 200-inch telescope was considered to be very risky, now primary mirrors with focal ratios of $f/1$ have been made. The faster primaries have resulted in smaller domes and stiffer structures.

Faster effective focal ratios with wide field correctors are becoming more popular to support the relatively small detector formats of CCD detectors and mosaics of CCD detectors. An example is the U.S. Naval Observatory's new 1.3-m telescope with a $f/2$ primary and an effective focal ratio of $f/4$. This telescope has a flat and corrector field 1.7 degrees across.

The dome seeing has improved to the point where the optical quality and collimation (alignment) of the optics have became significant. As a result, several telescopes have been built with a motorized remote-controlled 5-axis secondary mirror mount. The secondary mirror mount allows collimation adjustments (tip/tilt, X-Y translation, and focus) to be made from the control room. Now the telescope operator can quickly change the collimation in an attempt to improve the images.

CCD detectors have driven the development of the focus motion to a point where the focus can be automatically set for best focus for each individual filter while the filter is being changed. Values for focus motion resolution of the secondary mirror for various Cassegrain systems are given in the table.

The mirror substrates are mostly made from near zero expansion materials although the spin cast and fabricated lightweight borosilicate blanks are also used. These lightweight blanks have some thermal advantages.

We are beginning to see the addition of adaptive optics to telescopes. In some cases dynamically correcting the received wavefront has produced image quality nearly diffraction limited.

2.2   Detectors

The CCD detector is now almost universally used for imaging, photometry, and for spectrograph detectors. Photographic plates are still used for the very large format applications, but mosaics of CCDs are replacing plates. One of the limits to replacing plates is the very high cost of large format CCDs. A 2x4K CCD can cost $90,000. A mosaic CCD camera can easily cost two to three times as much as a moderate size telescope.

2.3   Pointing

Pointing has improved to the point where we expect the telescope to go to coordinates with a r.m.s. error of less than 30-arcsec. This improvement has led to the greatest increase in efficiency of any telescope development. The performance increase is entirely due to the ability of the computerized Telescope Control System (TCS) to model the telescope pointing. This modeling includes coordinate transforms between "mean" and "apparent", corrections for atmospheric refraction, polar misalignments, mechanical and optical non-perpendicularities, flexure of the telescope structure, and other repeatable errors.

When modernized, the present generation and older generation telescopes provide this level of performance. DFM Engineering, Inc. has built more than 70 new telescopes and retrofitted more than 18 older telescopes with a modern TCS and find these telescopes point better than 30-arcsec r.m.s.. Some actual pointing data is given in the reference.

Accurate telescope pointing coupled with an efficient slew to coordinates reduces the time required to move from one object to another. For example, the TCS can efficiently drive both axes of the telescope to the target coordinates simultaneously. The time to move from object to object is now measured in seconds rather than in minutes—an efficiency increase of 60 times.

2.4   Tracking

The open loop (without optical feedback) tracking of present telescopes is mechanically not much improved from the previous generation. The ability to calculate and apply tracking rates to correct for refraction and other factors does lead to better long-term tracking. Presently it is possible to keep a star within a 1-arcsec box for many minutes.

2.5   Controls

The use of digital computers and modern servo motor controllers has traded the mechanical complexity of the previous generation telescopes for software complexity. Software complexity has a very low reproduction cost while mechanical complexity does not. Now a single servomotor is used to power each axis of the telescope through its entire dynamic velocity range from track speed to slew speed. The maintenance of the telescope is highly reduced while the reliability is very high.

2.6   Structures

The faster Cassegrain primary mirrors have significantly reduced the swing radius of the Optical Tube Assembly (OTA). This factor coupled with a well-designed fork mount has reduced the inertia of the telescope allows higher natural frequencies to be developed. We have measured the locked rotor natural frequency of one of our satellite tracking telescopes to be 14-Hz. Such a responsive system allows the telescope to perform very fast and precise motions over a wide dynamic velocity range. The tracking is better than 1-arcsec up to the maximum speed of 5 degrees sec^{-1}.

Present telescopes are usually made with a friction drive system consisting of a small diameter roller driving a large diameter disk. Typically, the "gear" ratio can be as high as 24 to 1. This type of drive has very high stiffness, zero backlash, and produces very smooth tracking. The drive system is also very efficient resulting in very low heat dissipation at the telescope which improves the dome seeing. The typical power dissipation for a 1.3-m telescope at the telescope is less than 20-watts.

2.7   Support Functions

Control of the dome, shutter, and windscreen is typically incorporated into the TCS. In some specialized cases, the dome has been accurately servo controlled to allow the telescope to look through an optical window while the dome is sealed.

Most of the observatory support functions were located within the observatory building causing heat losses and degrading the images. Studies were made of the airflow within the dome and improvements were made to the dome ventilation including insulating the dome and even refrigerating the primary mirror and dome during the day. Sources of heat were removed from the dome. The dome seeing improved until images better than 1-arcsec became expected.

Some instrument support functions have been taken over by the TCS. These include changing filters and operating off-axis guider stages. However, incorporating instrument and data functions into the TCS is not generally done because the traditional telescope functions do not change over the life of the telescope while the instrument functions are constantly changing.

3.   FUTURE TELESCOPE PERFORMANCE REQUIREMENTS

The efficiency of collecting astronomical data has grown by many magnitudes. This increase in data gathering has been driven by advances in detectors and has resulted in the nightly collection of vast amounts of data. Fortunately, high-speed data transmission over the Internet is routine and almost free. Also, data processing has become much more efficient allowing more data within an image, for example, to be investigated creating even more data. The data may be partially pre-processed (pipelined) before transmission to reduce the bandwidth requirements.

The future performance of telescopes is being driven by the cost to operate the telescope. Part of this cost is the travel time and expenses to go to the observatory. Usually the observatory is located at an excellent astronomical site that is remotely located from the institution. Often data from different hemispheres is needed. In order to reduce the costs and the skill level, telescopes and instruments are starting to be operated remotely and robotically. Some systems are in operation now, and I expect many more systems will become remote or robotic in the next generation.

3.1   Optics

Because the telescope may be located at an excellent astronomical site, the optical quality will need to be sufficient to preserve the site's inherent image quality. Present optical quality is sufficient to meet this requirement.

The addition of adaptive optics to telescopes to dynamically correct the received wavefront will become more generally used as the cost for this hardware becomes more affordable. Correcting the wave front can produce nearly diffraction limited images.

3.2   Collimation

The optical collimation (alignment) will need to be very good to support the expected seeing limited images. This speaks highly for a 5-axis focus housing. The average user should not be allowed access to the collimation controls; they should be reserved for maintenance personnel only.

3.3   Pointing and Tracking

The pointing and tracking of the future telescope is discussed below in some detail. Both need to be better than the present generation telescopes requirements. Many of the present generation telescopes can meet these requirements or with some additions can meet the requirements.

3.4   Controls

The Telescope Control System (TCS) should probably reside in a computer separate from the computer that controls the observatory. This will simplify the development of instruments and the data collection and transmission.

A reliable means to initialize the telescope and instruments will be required. This may consist of absolute encoders, home switches, or zero reference marks (electronic fiducials). The software controlling the telescope will need to be able to perform self-initialization.

3.5   Structures

The present telescope structure is sufficient for the future telescope. A reliable means of initializing the telescope and instruments will be required.

3.6   Support

Because the telescope may be located at an excellent astronomical site, the observatory and telescope shelter will need to support the expected image quality. The observatory also will need to provide a safe environment for the telescope and instruments; for example, protection against inclement weather, provide security, and provide sufficient space for maintenance. A wide band communications link will also be essential.

Table 1.
Performance Summary for Past, Present, and Future

GENERATION	PAST	PRESENT	FUTURE	Units
Optical	0.5?	0.25	0.25	waves p-v
Quality	0.2?	0.05	0.05	waves r.m.s.
Pointing	1200?	30	20	arcsec r.m.s.
Tracking	2	1	1	arcsec r.m.s.
Focus				
Readout	25	2	2	microns
Positioning	50?	2	2	microns
Temp Comp.	some	many	all	

4. PERFORMANCE REQUIREMENTS FOR THE INTERNET TELESCOPE – AN EXAMPLE OF THE FUTURE GENERATION TELESCOPE

Table 1 presents a summary of the evolving nature of telescope performance requirements, and what we expect the needs of the next decade will be.

4.1 Definitions

Local: Local control means occurring from the control room. Most modern observatories are now operated from a control room (or warm room) not located in the same space as the telescope and instruments. The telescope and instruments are controlled by the operator interfacing to a computer.

Remote: Remote control means operating away from the control room. It may mean operating from the observatory floor, which is often done for public nights, for example. Remote may also mean operating or observing using dedicated cables from a distance such as from a planetarium hundreds or a few thousand feet away.

Far Remote: The proposed meaning of this phrase is to indicate operating or observing from a distance where dedicated cables are not used. For example, the observatory could be controlled over a campus Local Area Network (LAN).

Internet Access: The proposed meaning of this phrase is to indicate operating or observing from a far distance where the communications between the observatory and the user is performed over the Internet. The distance is indeterminate, the user may be anywhere in the world as long as they have access to the Inter Net.

Remote Observing: Observing from a location other than the control room or the observing floor. Also see Internet Access. This phrase has two distinct operational modes defined below.

Unattended Remote Observing; Attended Remote Observing: Remote observing places considerable demands upon the hardware. If all of these requirements are totally automated, the observing may be performed without human intervention (Unattended). If some of these requirements are performed by an attendant, then the phrase "Attended" is used.

Robotic Telescope: A robotic telescope accepts commands from another controller. Most modern professional and many amateur telescopes may be considered to be capable of being operated in a robotic manner.

Robotic Observing: This phrase usually means that the telescope and its instrument are being commanded to perform routine observations that have been preprogrammed. Such observations may be performed attended or unattended. One example of this system is the various seeing monitors we have built which measure the seeing and atmospheric extinction coefficients.

Scheduler: A scheduler is a software program that receives observing requests and either allows direct access to the telescope and instruments or sets up an observing queue based upon priorities, weather conditions and other preprogrammed factors.

Dynamic Scheduler: A dynamic scheduler allows observations in process to be interrupted by incoming requests of higher priority and resorts the queue to accommodate the new request.

4.2 The Internet Telescope

An Internet telescope allows a remote user to take command of the observatory and perform observations. Typically, we are discussing *real time direct imaging* rather than photometry or spectroscopy, although most of the performance requirements are similar for the various other instruments. We are not discussing a "batch processing" mode of operation where a set of observations are submitted well in advance of when the observations are performed. In the batch mode, a scheduler establishes when the observations are performed and afterwards, the data is sent to the user.

In the real time mode, the next scheduled remote user is recognized by the scheduler at the beginning of their time slot and is allowed remote access to the telescope. The user then specifies the coordinates of the first object of interest and the telescope slews to those coordinates. This object location may be specified by entering coordinates or by using a graphical user interface (GUI). The user then specifies how the imaging is to be performed by specifying the exposure time (integration time), the desired filter, and other camera parameters. The image is acquired and transmitted to the remote user and the next observation is performed. This sequence continues until the user's allocated time slot is completed and the next scheduled user is granted access to the observatory.

The control sequence becomes:

1. Recognizing the next remote user
2. Accepting the user's commands
3. Slewing the telescope to the object and tracking the object
4. Selecting the desired filter
5. Setting up the camera and controlling the integration
6. Transmitting the image data to the remote user
7. Begin the sequence again at step 2 until the allocated time expires
8. Go on to next remote user

The above sequence is sufficient to acquire images for relatively short integrations if the telescope points accurately enough to place the desired object near the center of the field of view of the camera, the focus is correct, and the tracking is good enough. If the object isn't centered well enough and or the focus isn't correct, then additional integrations would be required to properly adjust the telescope pointing and tracking, and adjust the imaging system. These corrections could use a considerable fraction of the observing time. To make the sequence efficient, user friendly, and reliable, the complete imaging system needs to be "transparent" to the remote user—a point and click operation.

The telescope functions are no different for the Internet telescope than they are for any telescope, but the performance requirements are more stringent. The telescope needs to point open loop (without feedback from the image) to a fraction of the field of view, the focus needs to be stable with temperature and accommodate any focus changes needed by different filters, and the telescope needs to track sufficiently well open loop to meet the imaging requirements.

4.3   Pointing

When a telescope is operated locally, if the object is not in the field of view or not well centered in the field of view, the operator can move the telescope using the hand paddle to center the object. The remote user may have this motion capability, but it is awkward and time consuming to perform these motions at the end of a relatively slow data transmission line. It is far more efficient if the telescope goes to coordinates and places the object near the center of the field of view.

Telescope pointing is usually characterized by specifying the r.m.s. value for the pointing error on the sky. An r.m.s. pointing error of about 1/4 of the field of view of the camera will usually place the object within the central 1/2 of the field of view. This value should be considered to be the absolute minimum pointing performance, as some objects will fall at the edge or outside of the field of view. An object falling outside of the field of view for a remote user should be considered to be an unacceptable error. A remotely operated telescope should probably point two or three times better or about 1/10 of the field of view to avoid this problem.

Table 2.
Required Pointing Performance

Aperture		Focal Length		Plate scale	Field size	Pointing
inches	*f*/ratio	inches	mm	arcsec mm^{-1}	arcmin	arcsec, r.m.s.
14	11	154	3910	53	8.8	53
16	10	160	4000	52	8.6	53
16	8	128	3250	64	10.6	64
20	10	200	5080	41	6.8	41
20	8	160	4000	52	8.6	52
24	10	240	6100	34	5.6	34
24	8	192	4880	43	7.1	43
32	10	320	8130	25	4.2	25
32	8	256	6500	32	5.3	32
40	8	320	8130	25	4.2	25
51	8	408	10363	20	3.3	20
51	4	204	5182	40	6.6	40

4.3.1 Typical Pointing Performance

The field of view is a function of the size of the detector and the focal length of the telescope. To image a large field of view, one wants a large detector and a short focal length. Unfortunately, large CCD detectors are very expensive, and short focal length, moderate to large aperture telescopes, are also expensive. Some practical examples follow using a commercially available CCD camera with a 10x10-mm CCD chip and our 1/10 of the field of view rule.

From Table 2, one can see that the recommended pointing performance is 20- to 60-arcsec r.m.s. (0.5- to 1-arcmin). The larger the telescope, the better the pointing performance needs to be. These values are "open loop" pointing, that is, with no optical feedback from the image.

With optical feedback, the required pointing becomes a function of the field of view for acquiring the image that may be smaller than the science imager field. There are some additional complications that are discussed later.

4.4   Initialization

Reasonably good time keeping and initialization hardware and software need to be part of the telescope control system to allow an auto-initialization procedure accurate enough to automatically find the first star. The time keeping should be good to one second. The initialization hardware (home sensors for example) needs to be good to about 15-arcsec.

4.5   Focus Stability

Efficient use of the telescope requires that the telescope focus remains within an acceptable tolerance during the entire night's operation over temperature changes and at different zenith distances. The temperature compensation may be accomplished passively, for example, by using Invar spacers, or by an active control. If an active focus temperature compensation is used, it probably should only adjust the focus when the telescope is not integrating. Any image motion due to focus movement should be very small– a few arcsec at most.

The focus mechanism needs to have sufficient motion resolution so that the smallest change in focus motion does not increase the image size significantly compared to the seeing disk. Also, the focus motion needs to be encoded at a resolution equal to or finer than the minimum focus motion. This is especially true when using a CCD camera, because the readout (and data transmission) may require considerable time. Time spent focusing the telescope reduces the efficiency of the system.

Each filter has an optimum focus position. It is advantageous to have a focus drive system that can automatically set the telescope focus position for each filter when the filters are selected.

If the remote user is allowed to change the focus position, then there needs to be an automatic return to the nominal focus position for each new user. This will prevent the last user from leaving the telescope in an unknown focus position.

The following Table 3 shows some examples of the required optical spacing change to produce 0.2-arcsec image enlargement due to defocus. The first two optical systems show a defocus effect that is very sensitive to the change in optical spacing due to the large amplification by the secondary mirror. The temperature change that produces 0.2-arcsec image enlargement in an uncompensated aluminum structure is shown in the fourth column.

Table 3
Optical Spacing Change for 0.2-arcsec Defocus

Telescope Size	Allowable Spacing Change		Allowable Temperature Change of Aluminum structure (°F)
Aperture (in)	Design of Optics	microns	
14	*f*/2-F/11	1.4	0.20
16	*f*/2-F/10	1.6	0.20
16	*f*/3-F/8	3.2	0.32
20	*f*/3-F/8	4.0	0.34
24	*f*/3-F/8	4.8	0.36

All of the above optical systems are very sensitive to temperature changes when fabricated from an uncompensated aluminum structure. A steel structure will be about half as sensitive. Some form of temperature compensation should be employed. A passive system using Invar spacers will be about 100 times less sensitive than the uncompensated aluminum structure. An active system will require frequent focus adjustments-perhaps as often as once every few seconds. If there is any image motion due to the focus motion, it will require a guiding system fast enough to remove the image motion. This will limit the ability of the autoguider to integrate to see faint guide stars.

Using an Invar spaced temperature compensated structure at a less than ideal site should allow the use of a *f*/3- *f*/8 telescope over a temperature range of about 50°F degrees without actively correcting the focus for temperature. It is possible to improve on the Invar spaced metering system using additional temperature compensating elements. Another approach is to use a carbon fiber reinforced composite structure. Such a structure can be made with a zero temperature coefficient at great effort and expense. Commercially available carbon fiber composite tubes are typically not as good as Invar spaced structures.

4.6   Tracking Performance

The ideal telescope would track for the entire imaging integration with an error too small to be seen in the image. Depending upon the site seeing, this may mean tracking to a fraction of an-arcsec. Even a "perfect" telescope can't track to a fraction of an-arcsec for more than a few seconds. The earth's atmospheric refraction at a modest zenith distance can easily change the Right Ascension tracking rate by 0.01-arcsec per second and introduce a Declination tracking rate of 0.01-arcsec per second. A 20 second integration will see a motion of more than 0.2-arcsec that is noticeable. The seeing will also introduce image motions of a similar or larger amount. Examples of the change in tracking rates due to the earth's atmospheric refraction at delination = 0° and 35° degrees latitude are given in Table 4.

Table 4.
Tracking Rate Corrections for Refraction

HOUR ANGLE (hours)	R.A. Rate Change (arcsec sec⁻¹)	DEC Rate (arcsec sec⁻¹)
-3.5	-0.007	-0.006
-4.0	-0.011	-0.011
-4.5	-0.023	-0.023
-5.0	-0.068	-0.069
-5.5	-0.182	-0.184

From this data, you can see that without correcting the tracking rates (R.A. and DEC.) at 4.5 hours you cannot track for more than about 50-seconds with a tracking error of 1-arcsec without guiding even with a perfect telescope.

The best tracking telescope we have experience with is the U.S. Naval Observatory 1.3-m $f/2.2$ - $f/4$ telescope at Flagstaff, Arizona. This telescope will keep a star image within a 1-arcsec box for more than 20-minutes open loop (no optical feedback). This telescope also has a fully temperature compensated optical tube assembly (OTA) structure so there is no optical spacing change due to temperature changes.

At less than ideal sites, a telescope that tracks very well may be satisfactory for short integrations of perhaps up to 1-minute in duration. However, in good seeing conditions, some form of autoguiding is necessary. This is one area where the technology has not been well developed for remote internet observing.

The telescope should be able to track open loop for many seconds with a tracking error of less than 1-arcsec. The tracking should also be very smooth so guiding corrections need to be applied only occasionally.

4.7   Tracking Rates

The telescope and control system should allow the remote user to input tracking rates in R.A. and Declination of up to a few arc minutes sec⁻¹. These rates would allow satellite tracking, for example. When a new remote user is granted access to the telescope, the tracking rates need to be reset to sidereal.

4.8   Guiding

Optical feedback from the telescope is needed to correct the telescope tracking when the integration time becomes more than a few seconds under good seeing conditions. With less than ideal conditions, a telescope that is well corrected may track open loop for tens or perhaps hundreds of seconds. Spectrographic or photometric instruments require a reduced level of tracking accuracy. Direct imaging requires tracking to a fraction of the seeing disk. In good seeing conditions this can mean guiding to 0.25-arcsec or better.

Direct imaging with high precision or long integration times really needs to be guided. The challenge is finding a suitable off axis guide star and positioning the guide probe and detector on the star. Once the guide star is located, the actual mechanism to determine the centroid of the guide star's image and provide guide corrections to the telescope control system is fairly straight forward.

The classical approach is to center the target object in the field of view of the imaging camera and then offset the guide probe to the coordinates of a previously selected guide star or to move the guide probe in a search pattern until a suitable guide star is found. Often suitable guide stars are selected prior to the observing session for each field to be imaged. The offsets are calculated and the guide probe is driven to the proper offset to acquire the guide star. With prior knowledge of where to look for the guide star, this approach is fairly efficient. Without prior knowledge, searching for a suitable guide star can consume considerable observing time-especially if the guide detector has a relatively small field of view, has to integrate for many seconds, or reads out the image slowly.

4.9   Auto Guiding

An Internet telescope needs to be simple to use. This implies that locating and using a guide star is totally automatic. We cannot expect the remote user to input the coordinates of the guide star with every desired field to be imaged. The selection of the guide star needs to be performed by the telescope or autoguider control system. Locating the guide star is another pointing problem. Typically, the field of view of the guide detector is considerably smaller than the main imager further increasing the need for accurate telescope pointing.

Under reasonable seeing conditions, the autoguider should be able to find a star with sufficient brightness to allow guiding with less than a few seconds integration. The telescope response to the auto guider should be sufficient to allow the telescope to track to a fraction of an arcsec.

5.   CONCLUSION

Telescope performance has made significant advances in the past two generations. The *efficiency* has increased by several orders of magnitude in an attempt to keep pace with the development of CCD detectors. The future will see only small incremental increases in the performance of the telescope but will see large changes in how the telescope is operated. No longer will the telescope be operated by skillful observers with much tribal knowledge.

Tribal knowledge of telescope and instrument operation has already been seriously eroded. The computer and its software have greatly simplified telescope operation allowing observers to obtain data with little understanding of telescope optics, controls, structures, and support functions.

Telescopes have traditionally been designed to have a useful lifetime of 100 years. Retrofitting new control systems to old telescopes has shown that they can be modernized to meet present and future performance requirements. Unfortunately, we are beginning to see our throw-away society exert an influence on even these long-living creations. Presently there are two 1.3-m telescopes that have been abandoned after completing their 5-year observing schedule.

Whether the future move to remote observing is good for astronomy or not remains to be seen. I believe some balance is needed. After all, when a lay person sees Saturn for the first time through a telescope eyepiece, they always say the same thing, "It really does have rings!". No amount of pictures and images has really convinced them of this fact.

6.   REFERENCES

Naval Observatory Telescope-A Generation 3 Telescope
    http://www.dfmengineering.com/news_naval_obsrv.html
Retrofitting Telescopes:
    http://www.dfmengineering.com/news_mt_megantic.html
Pointing:
    http://www.dfmengineering.com/news_eng_article_4.html

Chapter 21

The Potential of CCD-Based Amateur Spectroscopy

Alan Holmes
Santa Barbara Instrument Group
Santa Barbara, California USA

Abstract: With the wide availability of CCD imaging cameras in the amateur community has come the ability to do meaningful spectroscopic measurement projects. Amateurs can now perform high-resolution stellar spectroscopy with 2-Å or better resolution on stars down to 12th magnitude, and spectroscopy on diffuse objects down to 19th magnitude arcse^{-2} with five-minute exposures and commercially available mass-market telescopes.

Key words: amateur research, spectroscopy, instrumentation

1. INTRODUCTION

Amateur spectroscopy is an emerging discipline with great potential for expansion in the near future due to the wide availability of CCD cameras. In the past, amateur spectroscopy was film-based, and was limited by the low sensitivity of film and the tedium of exposing the film, and processing the negative. The amateur then had to confront the problem of getting the spectral information off the film, which required a film scanner, scanning microdensitometer or measuring engine of some sort. While film scanners are commonplace now, this was not the case even five years ago.

CCD cameras with excellent sensitivity became widely available starting in about 1992, and now have been sold in numbers in the thousands word-wide. The bulk of these cameras are purchased primarily for imaging, but many users eventually wish to participate in some sort of observing program with their instruments. CCD camera purchasers typically own either a large Schmidt-Cassegrain (SCT) telescope, with 25- to 40-cm aperture, or a quality 10- to 15-cm aperture refractor. CCD users themselves tend to be professionals who may either be a doctor, dentist, or engineer in their day job, but have little if any formal training in optics or astronomy.

T.D. Oswalt (ed.), The Future of Small Telescopes in the New Millennium, Vol. I, 257–269.

In this paper, I derive the performance and limitations of spectroscopy undertaken with amateur size instruments. I encourage others to devise research programs so enthusiastic amateurs with this capability can make a contribution to the field of astronomy. Also, most teaching colleges will be limited to similar equipment for their students.

1.1   Potential Research Areas for Amateurs

The following list describes the types of measurements that an amateur might pursue with a spectrograph.

- Stellar classification
- Monitoring of spectroscopic binaries
- Identification of elements
- Line identification in emission nebulae
- Measurement of temperature of emission nebulae
- Stellar radial velocity
- Red shifts of quasars and galaxies
- Measurement of light pollution in a geographic area

This list, which is by no means conclusive, gives an idea of the different specifications that an amateur might desire for the spectrograph. For measuring radial velocity, or identifying spectral lines in stellar spectra, the amateur needs high resolution. For measurement of emission nebula temperature, a good calibration at lower resolution is more important. For stellar classification of stars that haven't already been measured, or for galactic red shift, high sensitivity is necessary.

2.   SPECTROGRAPH DESIGNS

Here I discuss three types of spectrograph designs that have good applicability for amateur work. More complex designs, such as echelle spectrographs, are too complex to use and calibrate for the amateur at this time. The designs I describe are all used in first order.

2.1   Slitless Spectrograph Using a Grism

Many amateurs have experimented with spectroscopy using a Grism, a grating-prism combination before the focal plane. This is illustrated in Figure 1.

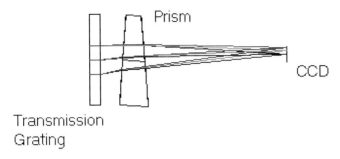

Figure 1. Slitless Spectrograph Optical Configuration

Light is preferentially diffracted into the first order on one side of a star's image, yielding a spectrum of the star. The simplicity of this technique is attractive for stellar spectroscopy, since locating and imaging the spectrum of faint objects is fairly easy. The type of image that is obtained is illustrated in Figure 2, showing a slitless spectrogram image of M-76, the Little Dumbbell. The negative attributes of this technique are the fact that the spectrum is convolved with the source, producing a complex pattern for extended objects. It also falls on a cluttered background, possibly other stars, and on a bright sky background, thus reducing the signal-to-noise ratio. For survey work this technique is excellent. It also works well for spectra of supernovae in distant galaxies, since the low dispersions used make classification possible.

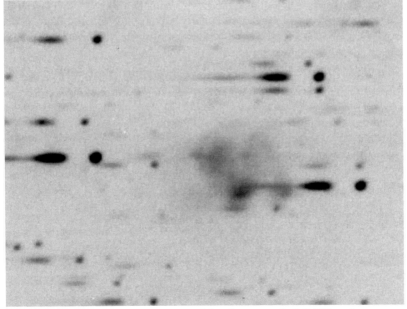

Figure 2. Slitless spectrogram of the planetary nebula M-76 region.

2.2   All Reflective Ebert-Fastie Design

This approach is the one used in the SBIG spectrometer, in folded form. It is illustrated in Figure 3. It is optically simple, and the all-reflective design produces well-focused spectra across the full width of a CCD, at any wavelength. The design has acceptable astigmatism if the f/ratio is slow, such as f/6.3 or f/10. The wavelengths are dispersed out of the plane of the paper.

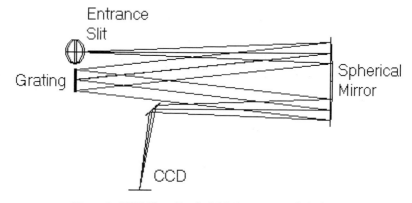

Figure 3. SBIG Ebert-Fastie folded spectrograph design.

The advantages are the compactness of the design, which is important when it has to be mounted on the telescope, and the high resolution possible. The biggest limitation is that the slit plane is imaged at one-to-one scale onto the CCD, forcing the slit to be small and reducing the sensitivity to extended objects. This will be explained in greater detail later. The fact that the plane of best focus is tipped at about 9° made it much easier to get light to the tracking CCD in the SBIG ST-7/8/9/10 camera line. The tracking CCD, in SBIG's implementation, shows both the entrance slit and the object to be analyzed. This greatly simplifies focusing of the telescope, positioning of the object on the slit, and tracking of a long exposure.

Figure 4 illustrates this approach. The tracking CCD image is shown on the left, with the planet Saturn positioned lengthwise on the slit. A portion of the spectrum is shown on the right, with the 6402-Å Neon line that was

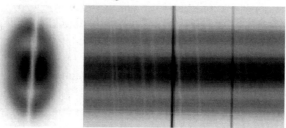

Figure 4. Spectrum of Saturn showing Doppler shift due to rotation (tilted vertical lines).

collected at the same time. Note the angle between the Neon line and the Fraunhofer lines in the planet's spectra. The shift is due to the Doppler effect and is a measure of the rotation speed and direction of the rings and the ball. This data was collected using a 60-sec exposure with an 11-in Schmidt-Cassegrain telescope.

2.3   Collimator-Camera Lens Technique with a Fiber Input

This technique is most similar to that employed by professional observatories. Light from a distant star or nebular region is captured by a fiber or fiber bundle and transmitted to a spectrograph off the mount, which can be large and heavy. Light from the fiber is collimated by an achromatic lens, dispersed by a first order grating, and reimaged onto the CCD by a camera lens specifically designed to be fast, such as *f*/1.4, and achromatic. The design is illustrated in Figure 5.

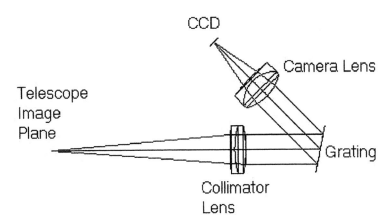

Figure 5. Collimator-camera lens configuration for fiber-fed spectrograph.

The design of the camera lens is quite challenging, since astronomical applications often require good performance deep into the blue, down to the calcium lines at 3944- and 3968-Å where the refractive index of most glasses is changing rapidly. This approach is very flexible since the grating can be easily changed, and different camera lenses used to achieve the dispersion desired on the CCD. The fiber tends to preserve the light cone of the illumination quite well, simplifying collimator design.

The problem of hitting the fiber tip with a star is significant. The simplest solution is to mount the fiber tip behind a beamsplitter such that some of the light goes to a CCD camera, and the bulk of it to the fiber. Both must be focused at infinity. Then one determines which CCD pixel corresponds to the location of the fiber tip by transmitting light out of the fiber tip, and using a corner cube in front of the telescope to reflect the light back to the CCD camera. The CCD camera electronics must then position the star on the correct pixel and hold it there by guiding the telescope.

The advantages of this technique are that it has high resolution and sensitivity, and gets the weight off the telescope. The major disadvantages are that it only takes the spectrum of one point on the sky at a time.

A variant of this approach is to use a smaller version on the telescope with a slit input. A similar procedure to determine the location of the slit on a guiding CCD camera's focal plane could be used. When performing radial velocity measurements one must worry about uniformly illuminating the slit, which is not trivial with a focused stellar image. If the slit is not uniformly illuminated it will shift the centroid position of a spectral line, which will produce an error of the velocity measurement. Narrow entrance slits minimize this effect.

3. PERFORMANCE LIMITS OF AMATEUR DESIGNS

In order to make some estimates of amateur capabilities, it is necessary to make some assumptions as to the constraints that apply to an amateur instrument. I will assume the CCD camera constraints summarized in Table 1 are applicable—most amateurs do not have routine access to liquid nitrogen for cooling a dewar, for example.

Table 1.
Amateur CCD Camera Performance

	SITE (Apogee AP-7)	Kodak (SBIG KAF-0401E)
CCD Parameter:		
Pixel size	24x24 microns	9x9 microns
QE at 6563 Angstroms	90%	64%
Read Noise	12 electrons	12 electrons
Dark Current	1 electrons/pixel/sec	0.25 electrons/pixel/sec
Sky Background (f/6.3)	38 electrons/second	3.0 electrons/second

I must also assume some constraints on the spectrograph. The main one is that gratings cannot be larger than 5-cm on a side before becoming prohibitively expensive, and are only available in groove frequencies of 150, 300, 600, 1200, and 1800 grooves mm^{-1}. A grating efficiency of 50% will be baselined. I will also assume that camera lenses are not available with less than 12.5-mm focal length since few CCD cameras can accommodate shorter lenses. For slitless spectrographs I will assume a grating frequency of 200 grooves mm^{-1} for a transmission grating since that corresponds to the popular

Rainbow Optics grating. For a telescope, calculations for an 11-in SCT operated at *f*/6.3 will be assumed, which is a median size and readily obtainable focal length in the amateur domain.

In general, the sky background will ultimately limit many kinds of spectroscopic observations. In Table 1 is included my best estimate of the sky background that will be obtained using the two common types of CCDs when used unfiltered. This estimate is based partly on measurement and partly on calculation, but I feel it is within ±20% of a typical dark site. Spectroscopically resolved measurements of the night sky brightness have been made from Kitt Peak at the zenith (Massey & Foltz 2000). The results are shown in Figures 6 and 7. The sky background in these figures is presented in terms of an equivalent brightness star in magnitudes arcsec^{-2}. I have recast the data in Figure 6 to units of photons sec^{-1} Å^{-1} for a CCD with a pixel size of 24x24 microns operating at the *f*/6.3 focus for an obscured SCT with a 15% area central obstruction and an spectrally flat optical transmission of 90%. This data is presented in Figure 8, and can be easily recalculated for other pixel sizes and *f*/ratios.

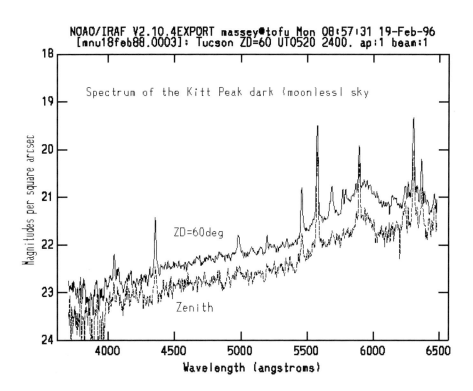

Figure 6. Zenith night sky spectrum at Kitt Peak in the visible region.

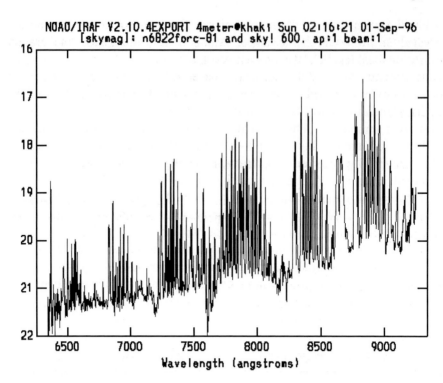

Figure 7. Night sky spectrum at Kitt Peak in the near infrared

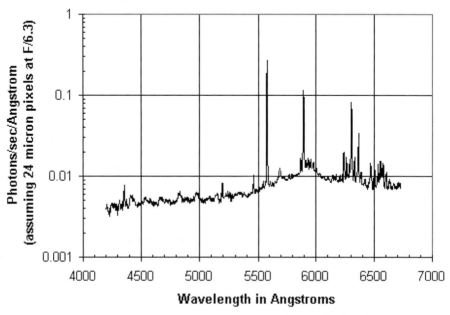

Figure 8. Kitt Peak sky spectrum recast to photons sec^{-1} Å^{-1} cm^{-1}.

I will quantify the performance of the three spectrograph options for wavelengths around H-α (6562.8-Å). At this wavelength the sky background is 0.01 photons sec^{-1} Å^{-1} for the assumed telescope configuration. A zero magnitude star of class A0 (Vega) has a baseline flux of 680 photons sec^{-1} Å^{-1} cm^{-1}. With the assumed 28-cm aperture telescope this is 3.2 x 10^5 photons sec^{-1} Å^{-1}.

3.1   Slitless Spectrograph Dispersion and Sensitivity

A slitless spectrograph using the Rainbow Optics grating placed 1 to 5-cm in front of the CCD would yield a dispersion from 1000 to 5000-Å mm^{-1}. Figure 9 shows the performance that would result. Vertical binning of the CCD is assumed to gather the charge over a 100-micron cross-dispersion region. The limiting magnitude is that magnitude where a signal-to-noise ratio (SNR) of 5 would result around 6563-Å.

Note that smaller pixels mean higher resolution. Longer exposures will improve the limiting magnitude, but only as the square root of the extra exposure. A 20-min exposure would be 1.5 magnitudes better if the CCD did not saturate on the background. In general, slitless spectrographs have very limited resolution, and sensitivity severely compromised by the background. CCD read noise and dark current are usually not significant—the sky dominates.

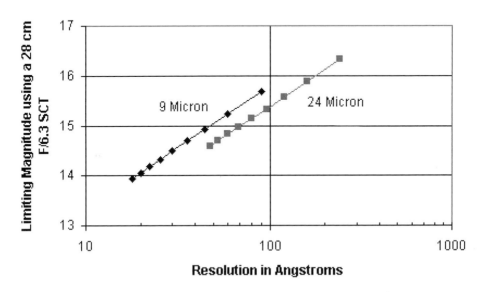

Figure 9. Limiting stellar magnitude using a slitless spectrograph and a 300-sec exposure.

3.2   All Reflective Ebert-Fastie design

The dispersion range of this design, based on the SBIG implementation with a 140-mm focal length spherical mirror (which is both camera and collimator), is 40- to 480-Å mm⁻¹. A new design with a spherical mirror from about 10- to 25-cm focal length could be considered which would accommodate other manufacturer's cameras, and still not be too bulky to ride on the telescope. The dispersion range practical with this approach then becomes 24- to 670-Å mm⁻¹. Figure 10 documents the performance against stellar magnitude. Figure 11 shows the limiting diffuse magnitude performance, which is more appropriate for extended objects. Note that here larger pixels seem to have an advantage, but this advantage comes at the expense of poorer angular resolution on the sky.

SBIG's spectrograph does not accept the full *f*/6.3 light cone—it is truncated to *f*/10 in one axis to allow the fold mirrors to be placed closer to the grating and reduce the astigmatism. This 30% light loss has been accounted for in the performance graphs. Also, all slit-based designs analyzed assume that the slit is two pixels wide when projected onto the CCD undispersed.

The Ebert-Fastie design has very attractive resolution in a small package but is limited in sensitivity to extended sources since the slit is imaged at 1:1 on the CCD.

3.3   Collimator-Camera Lens Technique

The collimator-camera lens technique has two significant advantages over the Ebert-Fastie. First of all, the nearly 4.5x focal reduction that can be achieved by using an *f*/1.4 camera lens dramatically helps the ability to measure a diffuse source. It also helps on a stellar source mainly because it is easier to get the light through the correspondingly larger slit. Secondly, a shorter focal length camera lens can be used, producing a lower dispersion on the CCD.

The demagnification present in this configuration produces a very significant diffuse magnitude sensitivity improvement (if the object size permits), and the lower dispersion possible (to 64-Å pixel⁻¹) is worth about 1.5 magnitudes for both stars and extended objects. Figure 12 shows the limiting magnitude vs. resolution relation for diffuse objects observed with the collimator-camera configuration.

Both the Ebert-Fastie and collimator-camera techniques can reach even closer to the sky background by using a larger slit and binning more pixels vertically, if the object is large in angular extent.

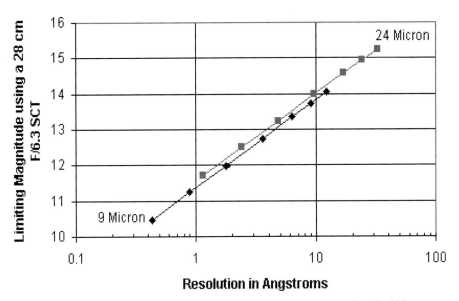

Figure 10. Limiting stellar magnitude using an Ebert-Fastie spectrograph and a 300-sec exposure. Relations for 9-micron and 24-micron pixels are plotted.

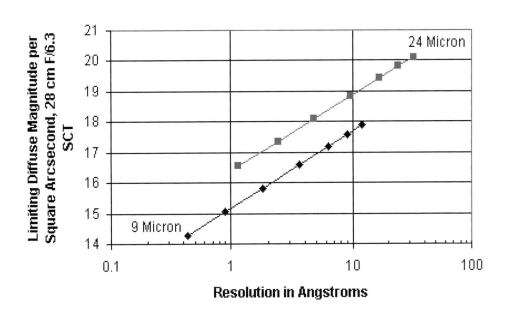

Figure 11. Limiting diffuse magnitude using an Ebert-Fastie Spectrograph and a 300-sec exposure. Relations for 9-micron and 24-micron pixels are plotted.

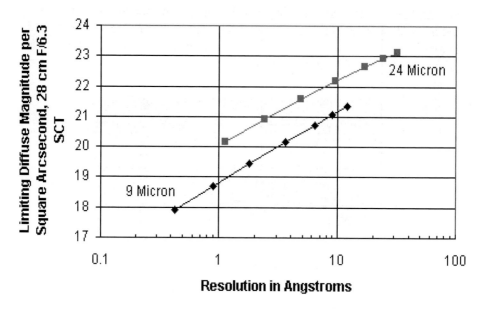

Figure 12. Limiting diffuse magnitude using a Camera-Collimator Spectrograph and a 300-sec exposure. Relations for 9-micron and 24-micron pixels are plotted.

3.4   Measurement of the Night Sky Spectrum for Light Pollution Characterization

The light pollution pervading American urban skies is becoming an issue recognized at the national level. It is clear from the night sky spectral data presented here that the instruments described can barely reach down to a brightness of 23rd magnitude arcsec^{-2} without some re-design. Fortunately, the sky has a large angular extent, making this detection possible. A simple version of the collimator-camera design that uses a much larger slit and more binning on the CCD can reach down to 23rd magnitude arcsec^{-2} levels even with the 9-micron pixel cameras with a SNR of about 5 at 6563-Å at a dispersion of 10-Å pixel^{-1}. This would nicely resolve both the shape and amplitude of the common mercury and sodium light pollution lines, providing a good baseline for governmental efforts to reduce this obscuring glow.

4.   CONCLUSIONS

Amateur equipment has reached a level where challenging measurements and observing programs can be considered. In Figure 13 is presented a last example of this: a measurement of the red shift of a galaxy (NGC 1275) made with an SBIG spectrograph using a wide slit, low dispersion (4.3-Å pixel⁻¹), and a 20-cm aperture telescope. The H-α emission from the 13th magnitude galaxy is shifted about 100-Å relative to the planetary nebula M76 in our galaxy, making for a definitive shift!

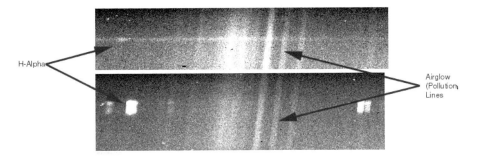

Figure 13. (top) Spectrum of NGC 1275 (Perseus A); (bottom) spectrum of M76.

5. REFERENCE

Massey, P. and Foltz, C. 2000, "The Spectrum of the Night Sky over Mount Hopkins and Kitt Peak: Changes over a Decade", PASP 112, 566.

Chapter 22

Small Radio Telescopes

Michael W. Castelaz, J. Donald Cline and Charles Osborne
Pisgah Astronomical Research Institute
Rosman, North Carolina USA

David A. Moffett
Furman University
Greenville, South Carolina USA

Abstract: Single antenna small radio telescopes, 30-m diameter and smaller, are often overshadowed by the few large dishes and multiple antennae arrays. Contemporary emphasis in radio astronomy is the construction of large antennae and multiple antennae arrays. Of course, the obvious reasons include greater flux sensitivity and improvement of spatial resolution. However, the relative youth of radio astronomy suggests that the usefulness of small radio telescopes have not yet been fully explored.

In this chapter several key challenges of observational and astrophysical interest are made to small radio telescope astronomers. The challenges include improvement in receiver technology, monitoring brown dwarfs, masers, and gamma ray bursts, surveys using low density interstellar medium indicator molecules, water maser surveys, and pulsar research. Also, as important as the astrophysics, is the continued education of future radio astronomers, and general public outreach. Hopefully, this chapter will inspire ideas for discovery using small radio telescopes.

Key words: radio astronomy, radio telescopes, antennae, receivers, brown dwarfs, masers, pulsars, interstellar medium, science education

1.   INTRODUCTION

Approximately 100 radio observatories exist in the world with both single antennae and interferometric arrays. If single and dual dish facilities are considered aside from arrays of three elements or more, then only 17 of

T.D. Oswalt (ed.), The Future of Small Telescopes in the New Millennium, Vol. I, 271–285.
© 2003 *Kluwer Academic Publishers. Printed in the Netherlands.*

75 radio telescopes have diameters 30-m and larger. Figure 1 shows the number of single and dual radio telescopes as a function of dish size. Radio telescopes with dishes less than 30-m in diameter in single or dual mode are defined in this chapter as small radio telescopes, or SRTs.

Radio astronomy is directly related to advances in physics, electrical and mechanical engineering, mathematics, and computer technology. Physics derives directly from the nature of celestial radio emission/absorption, including blackbody radiation, synchrotron radiation, molecular rotation bands, and electron spin. Electrical and mechanical engineering are obvious in the radiometers and massive radio antenna structures. Of course, the languages used include mathematics and the multiple dialects in computer technology. Indeed, an astronomer in the midst of radio research of some celestial phenomena must be skilled at each of these disciplines. Because radio astronomy combines so many disciplines, it is natural application vehicle for science and technology education.

Radio observatories using telescopes in arrays are also dominated by SRTs. Figure 2 shows the number of elements in arrays versus the diameters of the dishes in the arrays. None of the arrays use antennae larger than 25-m,

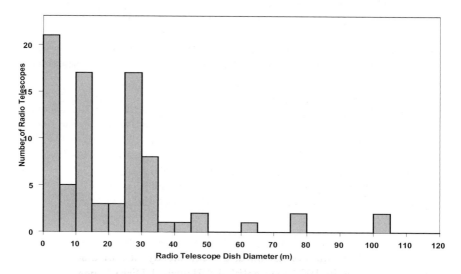

Figure 1. Histogram of radio telescopes throughout the world. The plot does not include Arecibo (305-m), interferometers of three or more elements, or VLBI programs. VLBI programs may use some of the radio telescopes included in the histogram or the hundreds of 5-m and smaller radio telescopes at universities or privately owned. The data was acquired via a survey of the Internet. Also, note that about 400 optical observatories are listed in the 2002 Astronomical Almanac, which implies at least as many optical telescopes. Optical telescopes dominate radio telescopes by a factor of at least four to one.

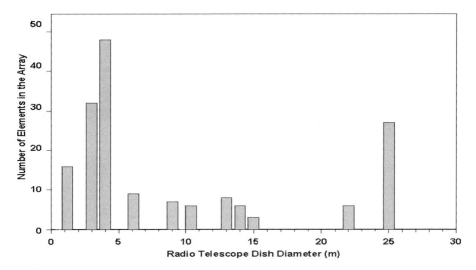

Figure 2. Arrays make up 18 of the radio telescope facilities in the world. The number of elements versus the size of the radio telescope dishes is shown. The Very Large Array (VLA), the largest dish diameter has 27 elements. Humain Radio Observatory has the largest number of elements with forty-eight 4-m dishes. The figure does not include the future NRAO Atacama Large Millimeter Array (ALMA) that will have sixty-four 12-m elements, nor the SETI Institute Allen Telescope Array which will have hundreds of 6-m antennae.

and most of the arrays use antennae smaller than about 10-m. Very long baseline interferometers, not shown in either Figure 1 or 2, use the radio telescopes included in the figures. Very large baseline interferometric (VLBI) programs are considered to be outside the definition of SRTs, and will not be discussed in this chapter.

A useful perspective of radio astronomy develops from a short history lesson. In 1930 Karl Guthe Jansky, a young physicist member of the technical staff at Bell Telephone Laboratories was assigned the task to study the arrival direction of atmospheric static at wavelengths of about 15 meters. These frequencies were used for ship-to-shore and trans-Atlantic communications. His 1932 paper described three distinct types of static: first; local thunderstorms; second; a steadier, weaker static due to combined effect of many distant storms; third, a weak steady hiss of unknown origin. His second paper in 1933 was entitled "Electrical Disturbances Apparently of Extraterrestrial Origin". This data showed a principal source maintaining a constant celestial position with a periodicity of 23-hr 56-min and a right ascension 18-hr, declination of 28° south or the center of our Galaxy. The sources of static located and described, Jansky never again worked on radio astronomy.

A lone radio engineer was the only person in the 1930's to attempt to make direct measurements to explore Jansky's observations. Using his own money and physical effort, Grote Reber began to construct equipment designed to record the cosmic static. Between June and September of 1937, he built a parabolic shape surface of sheet metal 31-ft diameter supported by a wood frame. From late 1937 through 1938 his UHF experiments detected no useful signals. In 1939, with a new 1.87-m (160 MHz) receiver, Reber detected emissions from the Milky Way. This work resulted in a paper in 1940. A second paper in 1944 contained the first map at radio frequencies showing the significant features of the sky from his location in Wheaton, Illinois.

Jan Oort made a prediction in 1944 that one should be able to find radio emission spectral lines. H. C. van de Hulst likewise predicted that a spectral line of interstellar hydrogen should be observable at 21-cm. WW-II started in Europe, shifting efforts toward radars and much higher radio frequencies. As a result of the war, radio hardware knowledge made giant strides. By 1950 Australian and British radio astronomers had collected detailed maps of bright radio sources. Howard Ewen and Edward Purcell made the first hydrogen emission line detection at Harvard in 1951. The observable neutral atomic hydrogen in interstellar space was a basic step in establishing a valuable method of understanding the structure of our Galaxy.

By 1960 the National Radio Astronomy Observatory (NRAO) was formed to develop and manage US radio observatories. The last 40 years of the 20th century continued the steady understanding of the total energy spectrum emitted within our universe.

We explore in this chapter several priority research areas and advances of radio telescope instrumentation that may evolve from these studies. Several key challenges of observational and astrophysical interest are made to small radio telescope astronomers. The challenges include improvement in receiver technology, monitoring brown dwarfs, masers, and gamma ray bursts, surveys using low density interstellar medium indicator molecules, water and methanol maser surveys, and pulsar research. Also, as important as the astrophysics, is the continued education of future radio astronomers, and general public outreach education.

2.   SMALL RADIO TELESCOPE TECHNOLOGY

Several key technology improvements will impact SRTs as astronomical research tools. The technology improvements are analogous to those seen in optical astronomy where the quantum efficiency of CCDs greatly increased the power of small optical telescopes. The effect of CCDs can be seen throughout discussion in other chapters of this book. Specific areas of SRT technology improvement include: Pseudomorphic High Electron Mobility

Transistors (PHEMT) low noise amplifiers, efficiency improvements in receiver electronics, universally accessible precision timing signals, embedded temperature controllers, multiple beam technology, and digital signal processing. One of the limiting factors in small radio telescope astronomy is the ability to accurately separate receiving system noise from the desired noise incident on the antenna from space. Unfortunately, the noise power collected is directly proportional to antenna collecting area. But fortunately, as the sensitivity of the system increases, proportionally less noise power results from the undesired contributors such as antenna sidelobes, spillover, feed losses, low-noise amplifier noise, and downconverter noise.

Pseudomorphic High Electron Mobility Transistors and Pseudomorphic Hetero-Junction Field Effect Transistors (PHJ-FETs) have reduced the noise temperatures of the critical first gain stages to 15-K at 4-GHz for *uncooled* devices. Twenty years ago this performance level would have been unattainable even with a US $50,000 helium cryogenic system. High quantity commercial satellite systems at 4- and 12-GHz have reduced the cost of the uncooled equivalent system components to less than US $100.

Improvements in receiver overall power conversion efficiency also resulted from the new PHEMT devices. This allows simple temperature control systems like Peltier coolers to be widely applied. The result is greater overall gain stability that minimizes system uncertainty about the source of noise power changes. Embedded controllers such as PIC or BASIC STAMP microprocessors greatly simplify precision localized temperature control.

In radio astronomy, and more particularly interferometry, access to extremely precise time and frequency standards has greatly improved with the worldwide availability of the Global Positioning System (GPS). In the 1960's atomic (Cesium or Rubidium) clocks were routinely couriered all over the world by calibration technicians to transfer and maintain system timing precision from one site to another. The GPS constellation of satellites allows nanosecond system timing precision with even simple receivers. This facilitates improved data quality and brings participation in interferometry networks within the realm of possibility for SRTs.

A recent development is multiple beam technology. Multiple feeds at different frequencies with duplicate backend electronics are placed in the focal plane of an antenna. A survey at multiple frequencies can then be done in much more quickly that a single frequency survey. Computer processing power improvements also facilitate this technology by keeping pace with increased real-time data flow.

Digital Signal Processing (DSP) adds a final significant improvement within the reach of SRTs. As processing speeds for personal computers doubled and redoubled beyond 3-GHz clock speed range, near-supercomputer processing speeds reached the desktop. Since the topic of

DSP and high clock speed computers is easily entitled to its own treatise, we simply state a series of examples of advantages of DSP and high clock speed.

- DSP allows often-unrealizable circuits, such as fractional Hz bandwidth filters, to be quickly created and tested in software.

- Interference can be dynamically subtracted from live signals where small gain variations can be compensated for in postprocessed data.

- Dispersion can be corrected on pulsar signals, making larger bandwidths an advantage instead of disadvantage.

- Massive amounts of data can be collected and processed in realtime reducing analysis time. Transient events can be trapped realtime and reacted upon, instead of being found months later in post processing.

3.   CHALLENGES FOR SMALL RADIO TELESCOPE ASTRONOMERS

As radio telescope technology improves, astronomers can pursue several areas of research with small radio telescopes. Some of the research requires mJy sensitivity, but not necessarily arcsecond or better spatial resolution. Monitoring, surveys, and high temporal resolution radio observations of stars and the interstellar medium are expected to lead to new discoveries. Education is equally important. The sections that follow present the challenges.

3.1   Variability

Small radio telescope facilities are often available for long-term studies, filling an important need for observations of variability. Brown dwarf radio flares, masers, RS CVn stars, and gamma-ray bursts signify the cutting edge astrophysical research studies of galactic and extragalactic variability.

Brown dwarfs are low mass stellar objects (<0.08 M⊙) that gravitational compression to initiate deuterium rather than hydrogen fusion. Brown dwarfs are low luminosity and cool with temperatures <1000K. The spectrum of a brown dwarf shows lithium (6708-Å) and methane bands, but does not show bands of TiO and VO and similar molecules seen in M and L stars.

Single brown dwarfs have been discovered as members of the Pleiades cluster (120 million years old) and several much older ones (~billion years old) within a few light-years from the Sun. Currently, only four brown dwarfs have been confirmed as companions to normal stars. Three of them are Gl 229 B, G196-3 B and Gl 570 D.

Because of the low mass, low temperature and low luminosity, there was no reason to think that brown dwarfs would be detectable at radio frequencies. Processes like synchrotron radiation require strong magnetic fields. Flares were not expected. However, brown dwarf radio flares were first reported by Berger et al. (2001a). The maximum flux was about 0.20-mJy at 8.5-GHz, an increase of 10,000 above the system noise. The observations were made at the VLA just at the right time, as the flares are unpredictable. Berger et al. (2001b) are conducting a search for radio emission from several nearby brown dwarfs using the VLA. They report that other brown dwarfs do not have emission as high as LP944-20 had. The mechanism for the 10,000 fold increase in flux is a mystery.

Considerable improvements in sensitivity are still needed for small radio telescopes to make similar measurements. Nevertheless, if the past 20 years are indicative of the rate of technology development, small radio telescopes may be able to detect similar fluxes in the near future. The SRT beam size is large for a single dish and localizing the flux would not be possible. However, a small radio telescope that detects a sudden increase in flux in the vicinity of a brown dwarf can report the discovery to a facility like the VLA, where confirmation and further measurements can be made. The theme of symbiosis between small and large radio telescopes is demonstrated.

A small radio telescope dedicated to monitoring brown dwarfs can be robotic. With a catalog of brown dwarfs, and proper scheduling, a telescope script would control the day-to-day pointing and flux measurement. A radio telescope aperture size on the order of 25-m would be ideal. At 1.4-GHz, the beam size is approximately 30-arcmin. Daily measurement would easily reveal extremely large changes in the flux, by a factor of 10,000, for example, within the beam. However, what is the probability of a single flare detection per year? If one brown dwarf per day flares, then at least 365 brown dwarfs would have to be monitored every day! So, a more important question is how often a brown dwarf will produce a radio flare.

An observing strategy to detect brown dwarfs must be developed to give a high probability of success without knowledge of brown dwarf radio flare frequency. Since brown dwarfs are found in open clusters, the wise course of observing would be to monitor several clusters every day. This has two advantages. First, a cluster may fall within a single beamwidth of, say, 35-arcmin. Any large increase in flux in the beam would then suggest a radio flare in the cluster. Notification to a radio observatory with high spatial resolution capability could then be notified of the event. The other advantage of cluster monitoring is that more than one brown dwarf can be measured at once, albeit the total flux from all the brown dwarfs in a cluster would be summed.

3.2   Masers

Microwave amplification by stimulated emission of radiation (MASERs) occur when electrons in molecules in the circumstellar shells of evolved stars, or in regions of star formation, are pumped into a metastable upper state by incident radiation. When the bound electron in the metastable state is hit by a photon of the right frequency, the electron can return to a lower state by emitting a photon of exactly the same frequency and direction as the incident photon which is scattered. The two photons move off precisely in phase. If each hits another electron in the same metastable state, there will be four photons in phase, and the process cascades. The 6.7-GHz transition of methanol is the second strongest centimeter masing transition of any molecule (after the 22-GHz water transition) and is commonly found toward star formation regions. It is typically stronger than 12.2-GHz methanol masers observed toward the same regions.

Masers vary. Water maser emission from W51M was observed to change from 1140-Jy to 152-Jy over the period of a month (Zhou & Zheng 2001). Prochter & Braatz (2000) analyzed archived data and made new observations of galactic water masers with the Haystack 37-m radio telescope. Variability on the order of hours was found. Rudnitskij et al. (2000) discovered water maser variability that is in phase with the optical light curve. The variability is thought to be due to periodic shocks driven by the stellar pulsation (Luttermoser 2000). Hydroxyl and Silicon Monoxide masers have also been found to vary in long period variable stars. In star formation regions, methanol and hydroxyl masers have been found to vary on time scales of a few months superimposed on longer variability on the scale of years (MacLeod & Gaylard 1996). One 12.2-GHz methanol maser varied from 50- to 300-Jy during a flare that lasted 35 days.

3.3   GRBs

Gamma Ray Bursts (GRBs) were discovered more than 30 years ago as energetic outbursts of gamma-rays. GRBs are extragalactic and are thought to be hypernova, a very massive star that explodes and collapses and forms a black hole.

In 1997, the satellite BeppoSAX provided rapid reports and somewhat precise positional information about GRBs to astronomers. BeppoSAX reported a GRB on May 8, 1997. Based on that report, Frail and Kulkarni used the VLA to detect radio emission coming from the BeppoSAX object a few days later. VLA observations of this and other GRBs are used to measure the size and speed of expansion of the GRB source. In addition, the VLA has been able to monitor the change in flux from a GRB for more than a year. "These observations indicate the extraordinary importance of radio astronomy for providing information that can be gained in no other way about one of the frontier areas of astrophysics," said Hugh Van Horn, Director of the National Science Foundation's Division of Astronomical Sciences.

Hence, radio emission associated with gamma ray bursts is important to measure for complete understanding of the energetic physics of these systems. The challenge for SRTs becomes the development of more sensitive and narrow band frequency receivers. SRTs can then be used to measure integrated variations in velocity structure. Seaton & Partridge (2001) interpret 3.6-cm observations of a GRB in 1989 and 1990 of about 1.7-mJy. Low radio frequency pulses may be detectable at few tens of MHz if GRBs are generated by relativistic strong magnetized winds (Usov & Katz 2000). Balsano (1999) operated a 74-MHz 1000-m^2 phased array to search for radio emission from GRBs. The telescope responded to within 10-sec of a satellite detection of a GRB. In their sample of 32 GRBs, two were detected co-temporally with the GRB. Upper limits on flux are a few kJy at 74-MHz.

3.4 Surveys

Small radio telescope facilities are often available for long-term studies, filling an important need for surveys. At 1.4-GHz a 25-m radio telescope has a 30-arcmin beam, whereas a 100-m dish has only a 6-arcmin beam. The smaller beam size of the larger dish is desirable for higher spatial resolution mapping, but time consuming for all-sky, or galactic plane, surveys. The galactic plane covers approximately 2000 deg^2 visible from the northern hemisphere. Mapping the galactic plane using a 30-arcmin beam requires at least 4000 pointed measurements, twice as many to oversample. Depending on the feed and receiver, and minimum flux requirements, a survey with a 25-m dish can take up to several years to complete. And, a similar survey would be prohibitive with a 100-m dish. Indeed, the beam size FWHM decreases as frequency increases. So, a survey of water masers at 22-GHZ, for example, would require ten times longer than a 1.4-GHz survey using a 25-m dish. At this point, the trade-off between limiting flux, and the size of the radio telescope aperture needs to be considered to make a timely survey at high frequencies.

Which surveys are important? Molecular surveys, such as CO, have been successful in discovering and studying the densities of regions of star formation. However, the peripheries of the dense molecular clouds are not well understood. In fact, CO as a tool of density measurement breaks down at the edges of molecular clouds where the optical thickness is small.

Instead, a tracer like CH, which can be used as a density measurement tool in low optical depth, becomes important (Magnani et al. 2000). Magnani et al. used the CH 3.335-GHz ground state hyperfine main line transition to map a small (40×40 pc) region at the Galactic Center, tracing the low-density molecular gas in that region.

Conducting a more general survey in CH with a single 25-m dish will sample the sky with a beam of 7-arcmin. Although small, known molecular clouds in the galactic plane can be mapped. With a 1-Jy receiver sensitivity, approximately 2 hours per data point is required. Hence, a 3-year survey could yield a complete survey of the galactic plane.

Maser surveys are also important. In the study of late-type stars, for example, Takaba et al. (2001) present results of a water maser survey of late-type stellar objects at 22.235-GHz using the Kashima 34-m radio telescope. They found the highest H_2O detection rate for the type of stars with a thin dust envelope (Mira/semiregular variables) among other types of sources.

Levine (1995) conducted a survey of 22-GHz water masers near the Galactic center and discovered ninety-nine sources. The masers are associated with stars and star-formation regions. These associations with star forming regions suggest that star formation may be common in the inner Galaxy.

A new and interesting diagnostic use of water masers has been proposed by Cosmovici, Pogrebenko, & Montebugnoli (1999) who predict 22-GHz water maser emission from planetary atmospheres under certain physical conditions. Water maser emission can be used for the search of extrasolar planets in planetary systems where comets are common.

3.5   Pulsars

Small radio telescope facilities are often available for long-term studies, filling an important need for pulsar astrophysics. Pulsars are neutron stars, survivors of supernova explosions responsible for the destruction of their progenitor stars. The progenitor is a short-lived massive star, greater than 3-$M_\odot$, converting its original core of hydrogen fuel into heavier elements until its core radiation pressure from fusion is overcome by the weight of upper atmospheric layers. Collapse occurs when the core contains iron. During collapse, the star's core is converted to neutrons, the atmosphere of the star is blown off at a tremendous velocity, and elements heavier than iron are created as escaping neutrons bombard atoms within the atmosphere producing rapid nucleosynthesis. In theory, the core is a neutron star 10-km in diameter, which is 1.4-$M_\odot$ remnant of the progenitor star. Typical neutron stars, with high surface bipolar magnetic fields from 10^8 to 10^{13} Gauss, rotate at rates between 1.5-ms to 10-sec.

Pulsed emission from neutron stars is due to the rotation of a neutron star's emitting region, either from the polar cap or its magnetosphere, through the observer's line-of-sight. The emission mechanism is coherent, yielding brightness temperatures in excess of $>10^{30}$ K. Although the mechanism is not yet known, two models are currently favored. One employs charged particles ripped from the surface accelerating along magnetic field lines in coherent 'bunches' (Ruderman and Sutherland 1975). The second model is a maser-like mechanism. The pumping mechanism is a plasma instability created as primary particles are torn and beamed away from the pulsar's polar caps and pass through a magnetosphere filled with secondary particles (Melrose 1995). This is akin to Type III solar bursts.

Most pulsars radiate only at radio frequencies, but a few have been discovered which radiate only X-rays and γ-rays, and a couple (Crab and Vela) emit radiation across the entire electromagnetic spectrum. Radio pulsars are easily observed from the Earth, while observations at other spectral bands require satellite telescopes. Radio pulsars are brighter at long wavelengths, fainter at shorter radio wavelengths, peaking in the 300-MHz to 1-GHz frequency range.

Large, sensitive telescopes (e.g., Arecibo 300-m, Parkes 64-m, Molonglo) have found over 1000 pulsars, a few of which are in the Magellanic Clouds. Because of their sensitivity, large radio telescopes are constrained to only perform searches, and monitor pulsars in exotic systems such as those in binaries and globular clusters. Due to observer oversubscription, only a percentage of large radio telescope time can be allocated towards pulsar research. On the other hand, SRTs, even with system temperatures in the hundreds of Kelvins, are capable of pulsar monitoring programs.

Small radio telescopes in Europe (Jodrell Bank, England; Torun, Poland), North America (Greenbank, West Virginia), Australia (Hobart, Tasmania), and South Africa (Hartebeesthoek) have monitored pulsars daily for over two decades. For example, Stinebring et al (2000) have monitored the 610-MHz radio flux density of twenty-one pulsars daily over a period of five years with a 25-m telescope at Greenbank, West Virginia. These small antennas are the workhorses of the pulsar astronomical community. SRTs monitor pulsar timing behavior, polarization, and observe the properties of the interstellar medium along the line of sight to pulsars.

The most significant challenge facing SRTs is recording and analysis of the measurements. A pulsar must be sampled with a frequency high enough to obtain a statistically significant number of data points (to overcome low sensitivity) over the short pulsar periods. Since pulsars have a small rotation period (1.5-ms to 10-sec) and rapidly varying flux within the pulse window, sampling has to be done on the order of millisecond to microsecond intervals. Current experiments underway at large telescopes are sampling at tens of nanoseconds. Since the sampling rate must be so fast, the input of the data from the receiver/detector must be accomplished with either a fast A/D system or a tape drive (with later processing).

Besides the detection challenge, the observer must also take into account dispersion time-smearing due to the frequency-dependent time-of-arrival of the pulses and time variable flux scintillation of the signal. To overcome the effects of dispersion, the observing band is usually split into many channels with a filter-bank. Then each channel is detected, time-shifted to the time-of-arrival of a reference frequency, and stacked with the other channels.

Small radio telescopes have been used to observe rapid changes in pulse periods. The pulsar's period is observed to slow down regularly, since a minute fraction of its angular kinetic energy is converted to radiation. However, a small population of young, energetic pulsars exhibit sudden changes in periodicity attributed to spin up and gradual slow down, also known as a 'glitch'. The internal structure of these stars must undergo a significant change of moment of inertia. So, constant monitoring is important for our understanding of this degenerate form of matter. The latest observed 'glitch' was detected by the University of Tasmania's 14m telescope, dedicated to observing the Vela pulsar 18 hours per day, every day (Dodson, McCulloch and Costa 2000).

Small radio telescopes also play a fundamental role in indirectly observing the interstellar medium along the line of sight toward pulsars they observe. Dispersion time-smearing of the pulsar signal can be used to determine the electron column density along the line-of-sight to the pulsar and/or estimate the distance. The same interstellar clouds of electrons in warm gas that cause the dispersion time-smearing also cause the radio emission to be scattered. The scattering acts on a fast and slow scintillation time-scale. Diffractive scintillation acts on the order of minutes to hours, and causes the pulsar's flux to fade and return much like how starlight 'twinkles' in the Earth's atmosphere. Refractive scintillation acts on time-scales of weeks to months. A very strange, refraction event was detected by the 25-m telescope at Greenbank at 610-MHz by Backer, Wong and Valanju (2000) toward the Crab pulsar. Observations of the pulsar profile formed by folding the time series onto itself at the pulsar period show that an 'echo' of the pulsar signal arose, then slowly merged with the primary signal over a period of weeks. When the echo disappeared, the Crab glitched.

Decades before the discovery of pulsars, Baade and Zwicky (1934) predicted the existence of neutron stars as a theoretical state of degenerate matter that might exist at the end of life of a massive star. The discovery of a pulsar by Hewish et al. (1967) was not tied to neutron stars and stellar evolution until one was found within the supernova remnant Crab Nebula by Staelin and Reifenstein (1968). Observations of pulsars have confirmed theory somewhat, but there are several which show extreme departures from theory. Without the constant monitoring using SRTs, the dynamic astrophysics of pulsars and their serendipidous events cannot be caught. Valuable information about the nature of pulsars and the interstellar medium would be missed.

3.6 Education

Education opportunities must not be overlooked. Small telescopes, radio and optical, are ideal for students to gain experience with instrumentation. Without understanding the equipment they are using, astronomers are at risk in their correct interpretation of astrophysical phenomena. Random and systematic noise events need to be identified and equipment troubleshooting needs to be justified.

Another risk the future generation of astronomers face is miscommunication with the engineers, telescope operators, and staff, of large or space-based telescope facilities. In order to correctly convey the instrument setup, astronomers must have experience with similar instrumentation. An astronomer who needs 1.4-GHz measurements must be fluent in bandwidth, calibration methods, sensitivity, spectrometers, feeds and receivers. Experience reinforces confidence in the method and quality of measurements.

Skills that undergraduate and graduate students acquire while making astronomical observations include astrophysics, electrical engineering, computer technology, mathematics, and chemistry. Small radio telescope observatories in collaboration with universities are ideal for teaching these skills. For example, graduate and upper-level undergraduate astrophysics and astronomy courses can be partially taught at the observatories. Visiting students and professors may bring their own instruments, use facility equipment, and participate in an ongoing survey or monitoring program.

Further, we cannot ignore pre-collegiate education. As a tool for fostering the excitement of science and discovery, small radio telescopes can be used effectively in middle and high school classrooms. The Radio JOVE, the Haystack Small Radio Telescope, and the Goldstone-Apple Valley Radio Telescope projects are three excellent examples of radio astronomy in pre-collegiate schools. Another example is the Pisgah Astronomical Research Institute's School of Galactic Radio Astronomy – an Internet classroom using the PARI 4.6-m radio telescope remotely. All of these promote astronomy, science and math skills, enhancing the excitement of research and discovery.

4.   SUMMARY

State-of-the-art radio telescope technology includes low noise amplifiers (PHEMTs), increased efficiency in receiver electronics, universally accessible precision timing signals, and digital signal processing.   As continued improvements in sensitivity are made, several challenges in astrophysical research can be pursued.

Long-term dedicated research is often possible with SRTs.   Hence, variability of masers, brown dwarf radio flares, and gamma-ray bursts are ideally suited for study with an SRT.   Continual observations of these objects are encouraged and worldwide campaigns providing maximum temporal coverage need to be developed. Only then can we fully learn about the dynamic universe.

The low density interstellar medium near regions of star formation can be surveyed using SRTs.   Indeed, an SRT is an ideal because of the match between beam size and frequency of observation. The beam is not too small, so large areas around regions of star formation can be mapped at frequencies of 3.3-GHz (frequency of emission of CH—a low density indicator). Water masers, observed with the smaller beams at 22-GHz, can still be mapped over large areas using an SRT. Of course, sensitivity limits the number of sources that can be observed.

A pulsar period will change.   However, high temporal resolution and long-term coverage is required. With access to time on an SRT, and with anticipated new technology in autocorrelators and receivers, significant progress can be made in understanding the last stages of stellar evolution of massive stars.

The next few years of radio astronomy promise revolutions in small radio telescope technology as they play an important role in probing astrophysical objects temporally and mapping their environments. SRTs are expected to experience the same rapid evolution as small optical telescopes discussed in other chapters in this book. We suggest the interested reader study the book "An Introduction to Radio Astronomy" by Burke & Graham-Smith (2002) to learn the fundamentals of radio astronomy.

5.   REFERENCES

Baade, W., and Zwicky, F. 1934, Proc. Nat. Acad. Sci., 20, 254
Backer, D. C., Wong, T., and Valanju. J. 2000, Astrophysical Journal, 543, 740, "A Plasma Prism Model for an Anomalous Dispersion Event in the Crab Pulsar"
Blasano, R. J. 1999, PhD Thesis. Princeton. "A Search for Radio Emission Coincident with Gamma-Ray Bursts"

Berger, E., Ball, S., Becker, K. M., Clarke, M., Frail, D. A., Fukuda, T. A., Hoffman, I. M., Mellon, R., Momjian, E., Murphy, N. W., Teng, S. H., Woodruff, T., Zauderer, B. A., and Zavala, R. T. 2001a, Nature, 410, 338-340, "Discovery of Radio Emission from the Brown Dwarf LP944-20"

Berger, E., Ball, S., Becker, K. M., Clarke, M., Frail, D. A., Fukuda, T. A., Hoffman, I. M., Mellon, R., Momjian, E., Murphy, N. W., Teng, S. H., Woodruff, T., Zauderer, B. A., and Zavala, R. T. 2001b, AAS, 198.6906, "Discovery of Radio Emission from the Brown Dwarf LP944-20 and Preliminary Results from an Ongoing Survey of Nearby Brown Dwarfs with the VLA"

Burke, B.F. & Graham-Smith, F. 2002, Cambridge University Press, "An Introduction to Radio Astronomy," 2nd edition.

Cosmovici, C., Pogrebenko, S., & Montebugnoli, S. 1999, Bioastronomy 99: A New Era in Bioastronomy. 6th Bioastronomy Meeting - Kohala Coast Hawaii - August 2-6, 1999. "The 22 GHz Water MASER Line: A New Diagnostic Tool For Extrasolar Planet Search"

Dodson, R. G., McCulloch, P. M., and Costa, M. E. 2000, IAU Circular 7347, "PSR 0833-45"

Hewish, A., Bell, S. J., Pilkington, J. D. H., Scott, P. F., Collins, R. A. 1968, Nature 217, 709

Hey, J.S. 1973, Neale Watson Academic Publications, Inc. NY, "The Evolution of Radio Astronomy."

Levin, D. A. 1995, PhD Thesis. UCLA. "A Survey of 22 Gigahertz Water Masers Within 2.25 Degrees of the Galactic Center"

Luttermoser, D. G. 2000, ApJ, 536, 923,"The Atmosphere of Mira Variables: A View with the Hubble Space Telescope"

MacLeod, G. C. & Gaylard, M. J. 1996, MNRAS, 280, 868. "Variable Hydroxyl and Methanol Masers in G 351.78-0.54"

Magnani, L.; Engebreth, B.; Hartmann, D.; Dame, T. M. 2000, Imaging at Radio through Submillimeter Wavelengths. Edited by Jeff Mangum. Publisher: The Astronomical Society of the Pacific, Conference Series, 2000. The conference was held June 6-8, 1999, in Tucson, Arizona. "Imaging the Galactic Center in CH: A Comparison with CO"

Melrose, D. B. 1995, Journal of Astrophysics and Astronomy, 16, 137, "The Models for Radio Emission from Pulsars - the Outstanding Issues"

Prochter, G. E. & Braatz, J. A. 2000, AAS, 197.4003, "Rapid Flux Variability of Galactic and Extragalactic Water Masers"

Ruderman, M. A., and Sutherland, P. G. 1975. Astrophysical Journal, 196, 51, "Theory of Pulsars - Polar Caps, Sparks, and Coherent Microwave Radiation"

Rudnitskij, G. M., Lekhet, E. E., Mendoza-Torres, J. E., Paqshchenko, M. I., & Berulis, I. I. 2000, A&AS, 146, 385. "Variability of the Water Maser Associated with U Orionis"

Seaton & Partridge 2001, PASP, 113, 6-9. "Possible Radio Afterglow of a 1989 Gamma-Ray Burst"

Staelin, D. H., and Reifenstein III, E. C. 1968, Science, 162, 1481

Stinebring, D.R., Smirnova, T. V., Hankins, T. H., Hovis, J. S., Kaspi, V. M., Kempner, J. C., Myers, E.; Nice, D. J. 2000, ApJ, 539, 300. "Five Years of Pulsar Flux Density Monitoring: Refractive Scintillation and the Interstellar Medium"

Takaba, H., Iwate, T., Miyaji, T., Deguchi, S. 2001PASJ, 53, 517. "Kashima 34-m Water Maser Survey of Late-Type Stars"

Usov V. & Katz, J. I. 2000, A&A, 364, 655. "Low Frequency Radio Pulses from Gamma-Ray Bursts" Zhou, J & Zheng, X. 2001, Ap&SS, 275, 431, "Short Time Variability of the Water Masers in W51M"

Chapter 23

Charting the Future of Small Telescopes
New Strategies for a New Decade

Terry D. Oswalt
Florida Institute of Technology and the SARA Observatory
Melbourne, Florida USA

Abstract: This paper examines some of the challenges faced by small observatories in the coming decade. Changes in planning scientific objectives, operations, personnel, funding, instrumentation, etc. can all help bring telescopes of aperture smaller than ~4-m into the 21st century as full partners with the world's largest telescopes. If they are to continue to play their essential roles in basic science in the 21st century, these new strategies must be developed and implemented now.

Key words: small telescopes, private observatories, national observatories

1.   A BIT OF HISTORY

Overshadowed by the 8- to 10-m class giants, and with 30-m class behemoths already on the drawing boards, it is generally assumed (and often lamented) that small telescopes will play a diminishing role in the 21st century. Here, and throughout these volumes, "small" arbitrarily is defined as ≤4-m, because it is this range that has been most frequently omitted from national and international planning activities, and most frequently subjected to the effects of limited operating budgets.

Diminishing availability, chronic underfunding, and the misconception that bigger is always better ("aperture envy"), may seriously limit the type of research and education initiatives that can be undertaken during the coming decade. Great increases in total photon-collecting aperture notwithstanding, the coming decade will see sharply *less* observing time per astronomer unless steps are taken now ensure that the world's suite of telescopes is properly funded, made widely available and that it includes a suitable range

T.D. Oswalt (ed.), The Future of Small Telescopes in the New Millennium, Vol. I, 287–300.

of sizes, instrumentation, and geographic locations. None of these goals can be met unless and until small observatories are included in national and international scientific planning and priority-setting activities.

In recent years these problems have been widely lamented in the U.S. and Europe, where closures and privatization have sharply cut the number of publicly available facilities. However, it is especially acute in the developing and economically depressed countries of the world, where national facilities may not exist at all and private facilities are rare. In these regions the next generation of astronomers is at special risk.

If small facilities are to play their essential roles in the 21st century, new models for operating them, setting scientific priorities, and funding them must be developed now. Another decade of procrastination will be too late, and the next generation of astronomers will pay the price of our short-sightedness. Let's look at some of the most pressing issues.

2.   SMALL TELESCOPES AND BIG SCIENCE

In the U.S. the First Annual Lowell Observatory Fall Workshop in 1996 catalyzed a community discussion of the essential roles played by small observatories. The consensus achieved there provided the impetus for several other meetings that dealt with related issues that could not be fully explored at Lowell, such as the preservation of small telescopes at national sites and the importance of small telescopes to high-level amateur-professional collaborations. Highlights of some of these meetings were summarized by Oswalt (1996, 1997, 1998).

Do small telescopes have a significant role to play in 21st century cutting edge astronomical research? That has been the focus of a number of workshops, spurred mainly by closures at several of the world's major observatories. Before deciding whether to close or to defend them, however, we need an answer to this question. Supported by the arguments given below, our main thesis is that *small telescopes do indeed perform essential functions that cannot be economically performed by larger instruments.* Some of these functions include:

Gateway— What are the chances that any proposal will receive competitively awarded time on an 8-m telescope before its main concept is proven, or before more than one type of peculiar object worthy of investigation is found? Slim to none! Fishing expeditions are rarely approved by 8-m telescope allocation committees. They simply aren't cost-effective. Small telescopes support the proof-of-concept studies and the fundamental observational surveys that make scientifically compelling observing proposals.

Observing Support— Ground-based and space-based multi-wavelength campaigns routinely need time- and/or location-specific coverage. Rarely do such campaigns require large telescopes. Also, discoveries made with large aperture telescopes often warrant long-term follow-up monitoring with small aperture telescopes.

Instrumentation— Small telescopes often provide the essential test-bed for instrumentation and detectors. Whenever possible, these time-consuming development efforts are best done on small telescopes before implementation on larger telescopes.

Special Projects— Long-term, synoptic and survey projects usually require many nights of telescope time and dedicated instrumentation. Rarely do they also require a large aperture. Many such projects can be supported for the cost of one large-aperture telescope. They usually provide the list of targets, and sometimes whole new classes of objects, for large aperture telescopes.

Development— New modes of scheduling, remote operation, automation, etc. are far less risky and far more cost-effective to evaluate on a small telescope. The ideas that work best should be exported to the larger telescopes.

Education— Clearly student access to small telescopes far exceeds access to large telescopes. Small telescopes are the recruiting and training ground for the next generation of observational astronomers and instrument builders, i.e. K-12 students, undergraduates, graduate students and post-docs. The importance of a strong public outreach effort is often overlooked. Small telescopes are the obvious focus for such activities.

The above is far from an exhaustive list, but it underscores the fact that the scientific productivity of even the present generation of 8-m class telescopes would be seriously compromised by a lack of many more small telescopes. The productivity, economic, scientific and political issues discussed by other chapters in these volumes add further weight to our premise that *small telescopes are absolutely essential to a healthy future for astronomy—even in the coming era of 30-m telescopes.*

3.   WHAT WE CAN DO

Instead of lamenting the neglect of small observatories in national scientific planning and funding decisions, let's focus on what astronomers can do during this new decade, not only to preserve and improve the productivity of our small facilities, but also to build new ones. Most of the following ideas and strategies could be applied anywhere in the world.

3.1   Networking

Probably the most serious impediment to the productivity of the world's small telescopes is that they are usually operated by small universities with very limited financial resources. Most have a limited pool of local users as well. Many are geographically isolated from the rest of the astronomical community. Under such constraints facilities tend to fall into disrepair and eventually close, or at least they become much less productive.

Collaborations can bring new users, instruments, upgrades, science and/or financial support to the small facilities around the world. The problem is finding a good match between a given facility's capabilities and the needs of potential collaborators. An easily accessible Internet registry or clearinghouse where those with projects and those with underutilized facilities can find appropriate matches would provide the necessary catalyst for such collaborations.

One of the first such registries was the North American Small Telescope Cooperative (NASTeC). It was founded by the late Jason Cardelli at Villanova University, and is now administered by the Southeastern Association for Research in Astronomy (SARA) as the International Small Telescope Cooperative (ISTeC[1]; see Henson 2002). ISTeC is a free no-obligation web-based resource for astronomers seeking collaborators and for potential collaborators seeking projects.

Especially if managed by an international organization such as the International Astronomical Union, an organization like ISTeC could be the main hub of a world-wide network of telescopes. The result, a new World-Wide Observatory (WWO), would nicely complement an international expansion of the National Virtual Observatory endorsed by the U.S. Decade Survey (see the chapter by McKee & Taylor 2002). The key to making it a reality is better communication and organization among observatories around the world.

3.2   Consortia

In the U.S. there are about a dozen consortia operating small observatories. The Southeastern Association for Research in Astronomy[2] (SARA; see Oswalt & Wood 2002) is an example of one way to guarantee a productive future for small telescopes otherwise threatened with closure. SARA assumed operation of a 0.9-m telescope at Kitt Peak National Observatory. This was one of the first U.S. national facilities to be 'privatized'. It was not the last. By the end of this decade it is likely that the

---

[1] Additional information can be found at the website http://astro.fit.edu/istec/

[2] See the SARA website at http://saraobservatory.org/

only publicly-accessible facilities available to U.S. astronomers will be ~4-m and larger. This alarming trend is by no means limited to the U.S.

Consortia also provide a means to develop *new* small facilities. The obvious benefits of consortia include stable financial resources, pooling of technical expertise, and improved opportunities for research and educational projects. Not so obvious are secondary benefits such as better student and faculty recruiting, as well as better regional networking between institutions. The importance of the latter is often overlooked—only one or two astronomers are typically found in the typical physics department.

Even in the U.S., where the economic and technical environment is rather favorable, a surprisingly small fraction of colleges and universities that employ astronomers are members of a consortium. Clearly there is an enormous untapped potential for new consortia-operated facilities.

3.3 Education

The world's small observatories are an ideal focus for projects that involve the exchange of students, teachers, university faculty, and high-ability amateurs. So far, only a few devote a significant fraction of their resources to these activities.

In the U.S. there are currently about 15 Research Experiences for Undergraduates (REU) programs supported by the U.S. National Science Foundation (NSF). A typical site, such as the one hosted by SARA[3], receives around 150 applications per year for 10-12 available internship positions. Opportunities for students outside the U.S. are even more limited. Clearly, even allowing for multiple applications, programs like these could be expanded several-fold and still not meet such strong demand. Why are there so few?

There's a catch. Education and outreach programs are nearly always a rather minor part of a national or international facility's mission. There are a limited number of such facilities and their staff is struggling to accommodate ever-tighter operational budgets. Moreover, faculty at large universities that operate comparable facilities are often indirectly penalized for participation in such programs, as it takes time away from their main research and proposal-writing efforts.

The faculties at predominantly undergraduate colleges and universities traditionally have strong commitments to and expertise in mentoring student/teacher research. Unfortunately, most do not have access to the requisite facilities for high caliber astronomical research. This is a problem that small university consortia can address.

---

[3] For information on the SARA REU program see http://astro.fit.edu/sara-reu/sara-reu.html

New consortia need to be actively fostered, and existing consortia encouraged to expand internship and outreach programs. This could be accomplished by creating competitively awarded funding programs that subsidize the operational expenses of research facilities of all sizes that are willing to undertake substantial education/outreach projects. Current programs in the U.S. support only expenses such as participant stipends, travel and lodging. The host institution contributes the operating expenses of the facilities used and the local academic resources needed by participants. Faculty mentors volunteer their own time and personal resources. These are substantial costs that outweigh the external funding provided by several times. This deters many institutions from participating. For most, it is more advantageous to seek conventional research grants.

How could more institutions be encouraged to engage in these activities? Perhaps small university consortia could administer some of the student/teacher research experience programs at national or international facilities. Partial operational support for small consortia or university observatories in exchange for public access is a way to preserve what we are losing as national sites close their small telescopes. In the U.S. several university consortia share access time with the astronomical community via partnerships with the NOAO. The Telescope System Instrumentation Program (TSIP; see the chapter by McKee & Taylor, 2002) ties public access to private facilities to federal funding for instrumentation projects. These ideas could be extended to include much more comprehensive student/teacher research, education and public outreach programs.

3.4   Funding

Small observatories often fall through the cracks in the funding process. There seems to be a common misconception that universities are able to fully support their operational expenses. Proof to the contrary is readily evident by the number of closures that have occurred during the last decade and the number that are woefully underutilized. Faculty who depend upon them are often hard-pressed to compete successfully for grants with those who have access to 8-m class facilities—"big science" is all to often equated with "big aperture". The future of small telescopes depends upon how successful we are in correcting these misconceptions.

In the U.S., a standard federally-funded "single-PI" grant for astronomical research ranges from ~$50-100K per year. The most common award duration is three years. Most funding agencies do not view the expenses of administering awards much less than this as cost-effective. Although a few private agencies such as the Research Corporation, Fund for Astrophysical Research, etc. and professional societies such as the American

Astronomical Society offer small grant opportunities, there are painfully few sources for projects that fall below the scope of these programs, i.e. in the $10-30K range.

The funding gap outlined above hits small observatories hardest. For example, a CCD camera or spectrograph for a 1-m class telescope often falls within this gap. The projects astronomers undertake with small telescopes frequently only need modest travel, publication, student salary and expendable supplies expenses. Even the annual operating expenses of a small observatory may fall in this range, especially if it is automated and web-accessible. Universities rarely are able to fill this gap on their own. There is a strong need for more small- and medium-sized grants programs.

The astronomer that depends upon small telescopes for the majority of his/her research also occasionally needs an 8-m class telescope. How does one support the travel and publication expenses related to the use of an international facility if it cannot be logically funded within a larger conventional grant proposal? It makes more sense to fund such observing runs through the facility that provides the time—on an as-needed basis. The telescope allocation committee has judged the proposal to have strong scientific merit; why subject the small project to a separate funding review?

3.5   Publications

The future of small telescopes, especially in the developing countries, depends most of all on effective communication. Periodicals such as Astronomy and Sky & Telescope contribute to public awareness of the importance of small telescopes around the world. However, a facility's survival depends upon its scientific productivity. Isolated or economically disadvantaged facilities often cannot afford subscription and page charges of conventional scientific journals. Consequently much data is never shared, collaborations are less frequently formed, and techniques and instruments are infrequently standardized. Isolated astronomers too frequently waste precious time and resources reinventing the wheel.

Few paper or electronic publications fill this need, which is strongest among small colleges and universities, in the states of the former Soviet Union, and in the developing countries. These are where the majority of the world's small telescopes are located. With about 1200 members world-wide, the IAPPP Communications[4] partly serves this niche, but there is much room for new venues by which archival data, small telescope instrumentation designs, faculty-student-amateur collaborative research results, and educational projects can be widely and cheaply disseminated.

---

[4] For information see http://www.iappp.vanderbilt.edu/

4. SMALL TELESCOPES AT INTERNATIONAL OBSERVATORIES

In the U.S. the majority of visiting observers at national observatories are from colleges and universities that do not have access to telescopes of ~1-m and larger. Public access to a well-equipped suite of optical and IR telescopes of all sizes was the single most important driving force behind the original university consortium that succeeded in convincing the NSF to fund the creation of Kitt Peak National Observatory over 40 years ago.

That core mission (Oertel 1989) has changed dramatically in the last decade, to one that emphasizes construction of large facilities. The most common reason advanced for this shift has been declining real-dollar operational budgets. However, as shown by many of the chapters in these volumes, the savings afforded by small telescope closures did not fund the present 8-m class telescopes and certainly further closures cannot fund the construction of the next generation of 30-m class telescopes. *This economic misconception is not limited to U.S. national observatories.*

Proponents of closing small nationally accessible telescopes have argued that the total area of aperture available to the world's astronomers has more than doubled since the new 8-m class telescopes became operational. What is seldom realized is that the number of hours of publicly-available observing time available to astronomers has decreased by an even larger fraction due to the closure of small facilities[5]. It's simple arithmetic: many small telescopes can support many astronomers' projects; a few large telescopes can support relatively few.

At the dawn of this new millennium, astronomers around the world are blessed with some of the world's most state-of-the-art facilities, but there may be precious little access time to go around. In an era of dwindling publicly accessible 1- to 4-m class telescopes, it will become more difficult to do the proper 'pilot studies' and 'follow-up' work necessary for maximal use of 8-m class telescopes. These points have been made repeatedly in these volumes. Here we will focus on some suggestions for increasing the role of national observatories in the coming decade in such a way that a full suite of telescopes can not only be preserved, but also enhanced.

The world's national observatories *can* regain a growing and cutting-edge role if steps are taken now. At least a few of the ideas outlined below merit further discussion among the funding-, planning- and policy-making bodies of the world's national observatories. It is especially important that we begin this discussion now, before all facilities smaller than ~4-m are only found on non-public sites.

---

[5] For a satirical look at the unchecked consequences of this rush towards the 'ultimate telescope' see Castelaz 1999.

Unquestionably, the national observatories' most serious problem is that much of the astronomical community has 'given up' on them. Many of us have conceded in the face of blue ribbon panel reports and national budget decisions that there is nothing we can do to reverse their decline—or at least to halt their transition to sites with only 8-m class telescopes.

If national observatories are to play a central role in the new millennium they must recoup the faith and support of the astronomical communities that they serve. They must re-engage their constituencies. Failing this, not only are national facilities at risk, but a growing number of universities which are now partners with them and/or tenants on federally-managed land are also at risk because they will lose the infrastructure that only a national observatory site can economically provide.

4.1 Improvements within a Zero-Sum Budget

Closure of telescopes and attrition of staff are the two most visible adjustments many of the world's national observatories have been forced to make during the last decade. This trend is likely to continue, *unless a more active role is taken by those who depend upon these facilities.* Here we outline what some of these actions might be. Most of the following suggestions focus on reductions of operational and personnel costs, but also on increasing the astronomical community's involvement in the operation and advancement of national facilities.

Closure vs. Renovation— While a case can be made for closing facilities that truly are beyond repair and renovation, most of the world's national facilities have few if any that are in this state. Most of the telescopes we now support could be brought into the 21st century as fully state-of-the art automated (i.e. maintained, but untended), single-purpose and/or queue-scheduled, Internet-accessible instruments, with improvements such as tip-tilt secondaries and active dome ventilation. The typical renovation and automation of a 1-m class telescope can cost as little as $10-20K, using mostly off-the-shelf standard hardware, software and installed by technical personnel already at these facilities, or in some cases by existing manufacturers of retrofit software and hardware.

Changes in permanent staff— Do national facilities always need a full resident staff of tenured Ph.D. astronomers? Why not bring staff from the astronomical community as 1, 2, or 3-year 'rotators,' the way some federal agencies staff other research centers and funding programs? Visiting scientists are much less expensive than permanent staff. Moreover, it would bring a sustained influx of new talent and ideas to these chronically under-staffed and under-funded facilities. Many could come on sabbatical appointments, subsidized by their home institution. Perhaps even more importantly, a staff that is well-populated by 'visiting scientists' would immediately reconnect the astronomical community to the mission and operations of national facilities. People are *always* more important than bricks and glass. Any reductions in permanent staff should be accomplished by normal attrition, and accompanied by *increases* in separately-funded visiting personnel from the astronomical community.

A larger role in graduate and post-graduate education— Some of the permanent staff and/or technical support positions at national facilities could be filled by advanced graduate students who are working on their theses at these facilities in exchange for administrative or technical duties. This is analogous to a university graduate or post-doctoral fellowship. Some or all of the costs could be borne by the home institutions of the students, or by federally-funded research fellowships. These programs would (1) provide much-needed support for graduate and post-doctoral fellows; (2) provide an innovative combination of research and education which is valued by most funding agencies; (3) provide an ideal training ground in observation and instrumentation; (4) decrease the number of permanent staff needed; and (5) give the next generation of astronomers an increased appreciation and stake in the health and welfare of their countries' national observatories. With the appropriate constraints, it might even be possible to allow universities to offer graduate-level observational and/or instrumentation courses at the national facilities in exchange for instrumentation, personnel or financial support.

Expansion of undergraduate research programs— Although small programs exist at some national sites around the world, there is much room for innovative expansion of academic year opportunities for students, summer opportunities for teachers, and many more summer internships for students. Nationwide classroom Internet access to the smaller telescopes would be easy—if they were automated.

Expansion of the public outreach effort— Most of the world's national observatories have a visitor center and public affairs office. Few have strong national public outreach programs, ties to regional public science centers, or strong educational programs in county primary and secondary schools. It's tough to do any of this in an era of declining federally-funded budgets. Why should national facilities be exclusively funded by federal money? Tap into city, county, or state funding sources for these education efforts (see Kuroda, 2002) for an excellent example of how research observatories in Japan often are strongly supported by city funds).

A national data archive— Since the advent of CCDs there has been a serious lack of systematic archiving of astronomical data. We now stand at risk of losing nearly two decades of data if steps are not taken to save it. In recognition of this, in the U.S., the recent Decade Survey has declared the establishment of a National Virtual Observatory (NVO) the highest priority 'small' astronomy initiative of the present decade (McKee & Taylor, 2002). Indeed, there is a growing call for a 'World Virtual Observatory' (WVO; see Djorgovski, 2002). The world's national observatories are uniquely positioned to provide the core of the WVO initiative. Within a particular country, the NVO/WVO could also support a funded archival research program similar to those NASA supports in the U.S. for such space missions as IUE, IRAS, HST, Chandra, etc.

More partnerships between universities and national observatories— In the U.S., operations of small national observatory telescopes have been successfully taken over by NOAO-university consortia and upgraded rather than closed (see Oswalt & Wood 2002). Universities provide cash for renovation and upgrades. The national observatory provides in-kind personnel and maintenance. This is a win-win solution for the astronomical community and new opportunities for such partnerships should be aggressively pursued. Note, however, it can be only part the remedy for declining federal support for the world's observatories, as it does not address the fact that astronomers who are not members of consortia still lack access.

Non-federal support for national facilities— In most cases, certainly within the U.S., national facilities are exclusively funded only through their federal operations agreements. Why constrain facilities that are already hard-pressed to be totally dependent upon the government funds for their annual budget? On a case-by-case basis why not allow them to compete for university, military, other federal agency, educational and commercial grants and contracts that are clearly related to astronomy and within the resources available at the facility? Those national facilities within the NASA and ESA countries could and should provide ground-support for space missions. Time earmarked for guaranteed ground-support observations should be paid for on an at-cost basis by the agencies or governments receiving it. This would subsidize tight observatory budgets and foster inter-agency and international collaborations.

4.2 Changing the Structure and Mission of National Facilities

Most national observatories were created with a mission to provide a full suite of state-of-the-art instruments to the astronomical community. Access to these facilities is usually via a merit-based allocation system. Some allow international applicants; others limit access to astronomers within the facility-supporting country or participating countries.

Over the past decade, the missions of most of these observatories have willingly or unwillingly evolved to focus only on *state-of-the-art large telescopes*. As many of the articles in these volumes have pointed out, this has not been driven primarily by dwindling budgets, or the need to fund the next generation of larger telescopes. How can the position of the world's national observatories be strengthened and their small telescopes preserved?

Restructure the governing boards of national facilities— In many cases the governing body of a national facility consists of scientists from relatively large institutions or consortia that operate other ground- or space-based facilities. Often these institutions' own facilities directly or indirectly compete with the national observatories for funding. Even with the most altruistic of motivations, this creates a difficult conflict of interest. Such conflicts are not allowed within funding agencies; why are they common among the governing bodies of national facilities? These boards need to be staffed by impartial users of the facilities they represent (see next suggestion).

'Associate membership' for governing boards— At least some of the governing bodies of the world's public observatories could include elected representatives from blocks of small institutions that do not have access to competing facilities and that depend strongly upon the national observatories for scientific research. Such representation could be delineated by region, by consortium or by institutional type.

Tenant services— Most of the world's national observatories share their sites with other countries', university's or consortia's facilities. Usually, tenants are charged a usage fee for sharing the infrastructure provided by the national observatory. Rarely is direct technical support provided to tenants. In many cases it would be possible to supplement the national observatories' operating budget by providing as-needed technical support of tenant facilities on a full cost recovery basis.

University and national observatory instrumentation initiatives— The world's national observatories need not be the primary development centers for instrumentation. Contracts between national observatory and university instrumentation groups can support the development of community-accessible instruments. This has occurred on occasion (e.g. the TSIP initiative mentioned by McKee & Taylor, 2002), but there are many more opportunities including, for example, agreements to upgrade and equip small publicly accessible telescopes at national sites.

International centers for software development— In the U.S., NOAO has already established the Image Analysis and Reduction Facility (IRAF) software package as the astronomical community standard for optical and infrared image processing. It is time to develop the next generation of standardized software that can support both the national facilities and other facilities in the astronomical community. Examples include software for automation, queue scheduling, data acquisition and handling, generic instrumentation drivers, training and simulation, etc. Educational software could also be included in this initiative. As for the NVO/WVO data archive, the national observatories could serve as a clearinghouse and documentation source for software that was originally developed elsewhere.

Centralized telescope scheduling— In the U.S., NOAO provides observing proposal solicitation and processing for several public and private observatories. This is a model that could be expanded further at NOAO, and to other public and private observatories around the world. It makes good sense to centralize and standardize the solicitation, review and scheduling process of telescopes as widely as possible.

Publications, public relations and graphic arts services— For many years the Astrophysical Journal was edited from offices provided by NOAO in the U.S. Perhaps other journals and astronomical book publishing projects could cost-effectively be located at national observatories' administrative offices, thereby subsidizing infrastructure costs.

Design and engineering services— Why not allow national observatories to contract out to the astronomical community whatever optical, mechanical, electronic design and engineering services it can? This would help keep permanent technical support personnel employed and drafting/machine shops fully utilized, at no cost to federal budgets.

Funding support for visiting observers— Years ago, in a better budgetary climate, visiting observers at national observatories received a full travel subsidy, as well as publication page charges for papers involving use of data collected at their facilities. In addition, travel support was readily available for students who accompanied their faculty mentors. Now, in order to pursue a research project an astronomer must win a research grant to provide the funding, and winning observing proposal(s) for access to the necessary facility(ies). In the increasingly competitive scientific environment, it is common to win one, but not the other. National observatories should subsidize the expenses of research projects that win observing time, but are not supported by other sources. This should be accomplished by increments to, not at the expense of, facilities' operational budgets.

5. CONCLUDING REMARKS

The users of small telescopes are most often students and faculty at small universities. In some cases they are highly competent amateurs. In the preceding discussion one recurrent theme has been that this single largest segment of the astronomical community has rarely been directly engaged in the process of charting the future of the facilities upon which they depend most. Clearly, the planning process must be more inclusive than it has been. Moreover, users must assume more responsibility for the operation, management and funding of their facilities.

Quite a few of the ideas in this chapter have been suggested elsewhere. However, it does not appear that concerted attempts have been made to integrate many of them into the plan for the national "telescope system" envisioned here in the U.S., nor elsewhere around the world.

International organizations such as the IAU can play an essential role in evaluating the options and making recommendations regarding the future of small telescopes around the world. The relevant IAU Commissions could:

- Define the problems most commonly faced by the world's small observatories at one of its triennial meetings, and endorse the most effective strategies by which they may be solved;
- Coordinate the development of an international network of small telescopes like ISTeC;
- Evaluate and coordinate the standardization of astronomical hardware and software;
- Foster "small grants" programs for small telescope instrumentation, exchange of students and faculty, and internships;
- Provide essential advocates for small telescopes to funding agencies and national planning committees.

The above is far from an exhaustive discussion of the challenges and potential responses to the problem of preserving the future of small telescopes. We hope to spur a discussion that leads to a productive future for telescopes of all sizes. Which can we develop and implement, together?

6.   REFERENCES

Abt, H. 2002, in volume 1 of this Kluwer series.
Castelaz, M. 1999, IAPPPC
Djorgovski, G. 2002, in volume 1 of this Kluwer series.
Henson, G. 2002, in volume 1 of this Kluwer series.
Kuroda, T. 2002, in volume 1 of this Kluwer series.
McKee, C. & Taylor, J. 2002, in volume 1 of this Kluwer series.
Oertel, G. 1989, private communication.
Oswalt, T.D. 1996, IAPPP Comm. 65, 40-42.
Oswalt, T.D. 1997, IAPPP Comm. 69, 1-5.
Oswalt, T.D. 1998, European Astronomical Society Newsletter 16, 6-8.
Oswalt, T.D. & Wood, M.A. 2002, in volume 2 of this Kluwer series.
Querci, F.R. & Querci, M. 1997, IAPPP Comm. 69, 6-8.

Index

316